STUDY GUIDE

for the Sixth Edition of

FIRE INSPECTION

AND

CODE ENFORCEMENT

♦ **Developed by John Joerschke**

♦ **Edited by Cindy Pickering**

Published by

Fire Protection Publications Oklahoma State University Stillwater, Oklahoma

The International Fire Service Training Association (IFSTA) was established in 1934 as a "nonprofit educational association of fire fighting personnel who are dedicated to upgrading fire fighting techniques and safety through training." To carry out the mission of IFSTA, Fire Protection Publications was established as an entity of Oklahoma State University. Fire Protection Publications' primary function is to publish and disseminate training texts as proposed and validated by IFSTA. As a secondary function, Fire Protection Publications researches, acquires, produces, and markets high-quality learning and teaching aids as consistent with IFSTA's mission.

Layout and design by John Joerschke and Associates, Stillwater, Oklahoma

Copyright © 1998 by the Board of Regents, Oklahoma State University

All Rights reserved. No part of this publication may be reproduced without prior written permission from the publisher.

ISBN 0-87939-159-6

Second Edition

10 9 8 7 6 5   Printed in the United States of America

If you need additional information concerning our organization or assistance with manual orders, contact:
    Customer Service, Fire Protection Publications, Oklahoma State University
    930 N. Willis, Stillwater, OK 74078-8045
    1-800-654-4055   FAX: 405-744-8204
For assistance with training materials, to recommend material for inclusion in an IFSTA manual, or to ask questions on manual content, contact:
    Editorial Department, Fire Protection Publications, Oklahoma State University
    930 N. Willis, Stillwater, OK 74078-8045
    405-744-4111   FAX: 405-744-4112   E-mail: editors@osufpp.org

Table of Contents

Preface

This study guide is designed to help the reader understand and remember the material presented in **Fire Inspection and Code Enforcement**, sixth edition. It identifies important information and concepts from each chapter and provides questions to help the reader study and retain this information. In addition, the study guide serves as an excellent resource for individuals preparing for certification or promotional examinations.

Much time and effort go into the design, development, layout, and printing of any publication. This **Study Guide for the Sixth Edition of Fire Inspection and Code Enforcement** is no exception. I would like to give special recognition to John Joerschke and the following members of the Fire Protection Publications staff whose contributions made possible the technical accuracy and visual appeal of this publication.

Barbara Adams, Associate Editor
Susan S. Walker, Instructional Development Coordinator
Mike Wieder, Senior Publications Editor
Don Davis, Publications Production Coordinator
Desa Porter, Senior Graphic Designer

Cindy Pickering
Curriculum Specialist

How to Use this Book

This study guide is developed to be used in conjunction with and as a supplement to the sixth edition of the IFSTA manual, **Fire Inspection and Code Enforcement.** The questions in this guide are designed to help you remember information and to make you think—they are *not* intended to trick or mislead you. To derive the maximum learning experience from these materials, use the following procedure:

Step 1:   Read one chapter at a time in the **Fire Inspection and Code Enforcement** manual. After reading the chapter, underline or highlight important terms, topics, and subject matter in that chapter.

Step 2:   Open the study guide to the corresponding chapter. Answer all of the questions in the study guide for that chapter. You may have to refer to a dictionary or the glossary in the **Fire Service Orientation and Terminology (O&T)** manual for terms that appear in context but are not defined. After you have defined terms and answered all questions possible, check your answers with those in the answer section at the end of the study guide.

✓ **Note:** *Do not* answer each question and then immediately check the answer for the correct response.

If you find that you have answered any question incorrectly, find the explanation of the answer in the **Fire Inspection and Code Enforcement** manual. The number in parentheses after each answer in the answer section identifies the page on which the answer or term can be found. Correct any incorrect answers, and review material that was answered incorrectly.

Step 3:   Go to the next chapter of the manual and repeat Steps 1 and 2.

Chapter 1 Authority, Responsibilities, and Organization

Matching

A. Match to their definitions terms associated with fire inspectors' authority, responsibilities, and organization. Definitions are continued on the next page.

_F_ 1. Person who holds a sworn public position

_C_ 2. Action that a fire inspector considers necessary to fulfill his/her responsibilities

_I_ 3. Manner in which a fire inspector carries out or performs an act or policy

_M_ 4. To assume the responsibility in total for any claims against an individual fire inspector

_A_ 5. A fire inspector's responsibility when a person has moved from a position of safety to a position of danger because he/she relied on the inspector's expertise

_L_ 6. Requirement that codes must be applied equally, within reason, to all applicable occupancies in a jurisdiction

_J_ 7. A ruling that must be applied the same way with all future code enforcement

_B_ 8. Adoption of regulations by a local jurisdiction following state/provincial laws exactly as drawn

_K_ 9. Adoption of regulations by a local jurisdiction using state/provincial laws as a basis but modifying according to local needs

_D_ 10. Code published by an independent agency such as NFPA or BOCA

_H_ 11. Code used to regulate activities within a building once it has been constructed

a. Special duty

b. Adopt by reference

c. Discretionary act

d. Model code

e. Responsibility to enforce

f. Public officer

g. Permit

h. Fire prevention code

i. Ministerial action

j. General variance

k. Enabling act

l. Duty to inspect

m. Indemnify

G   12. Official document that grants a property
        owner permission to perform a specific
        activity

True/False

B.   **Write *TRUE* or *FALSE* before each of the following statements. Correct those statements that are false.**

T   1. Fire prevention inspections are the fire service's most important nonemergency activity.

_____

_____

F   2. Only designated *fire inspectors* may perform fire inspections.

_____

_____

F   3. Fire inspectors can usually assume that statutes, codes, and regulations are essentially the same from one jurisdiction to another.

_____

_____

F   4. Fire inspectors who have the authority to issue a summons or write a ticket do not need law enforcement training unless they are likely to be involved in prosecuting fire code violators.

_____

_____

T   5. When discrepancies exist between internal and external inspections, the inspector should be satisfied as long as the situation meets or exceeds local code requirements.

_____

_____

T ~~F~~   6. Fire inspectors should avoid at all cost any action that might involve the jurisdiction in a civil rights case.

_____

_____

F   7. A jurisdiction may not legally contract with a private firm to attain expert assistance with code enforcement issues.

_____

_____

T   8. Property owners may legally refuse admittance to an inspector who has not attained a proper warrant.

_____

_____

T   9. State or local codes may state their own guidelines on right of entry by public personnel.

_____

_____

F ~~T~~   10. In most jurisdictions, fire inspectors have the authority to approve modifications of code requirements.

_____

_____

T   11. Depending on local statutes, noncompliance with the code may be either a criminal or civil violation.

_____

_____

F ~~T~~   12. All complaints from concerned citizens should be immediately forwarded to the appropriate fire department official for further action.

_____

_____

F   13. Fire inspectors must give the responsible party advance notice before investigating a complaint.

_____

_____

T   14. Fire inspectors should keep current on building, plumbing, and electrical codes.

_____

_____

F   15. Exchanging information with other agencies creates unacceptable and possibly illegal conflicts of interest.

_____

_____

T   16. Fire inspectors should monitor business license applications and business and occupancy permits issued.

_____

_____

F ~~T~~   17. The local fire inspector is responsible for enforcing federal regulations.

_____

_____

T   18. Federally owned buildings are not required to comply with local codes.

_____

_____

_T_ 19. The fire department should perform prefire planning visits and company walk-throughs at federal facilities.

_____

_____

_T_ 20. Fire inspectors should be involved in updating laws that apply to their fire inspection duties.

_____

_____

_F_ 21. Permits generally allow property owners to disregard or exceed code requirements.

_____

_____

_T_ 22. According to the model codes, bowling pin refinishing, insecticidal fogging, and parade floats are all subject to permits.

_____

_____

_T F_ 23. A permit authorizes the fire inspector's entry at any time to ensure compliance with code requirements and conditions of the permit.

_____

_____

_T F_ 24. If a permit application appears to be in order, the fire inspector may issue the permit immediately or perform a site visit to further investigate the request.

_____

_____

_____F_____ 25. A permit must explicitly state the actions it allows, its guidelines, and its budget.

_____

_____

_____T_____ 26. Each jurisdiction must have procedures for granting extensions for permits.

_____

_____

_____T_____ 27. If an inspection shows that false statements or misrepresentations were made on a permit application, a permit may be revoked.

_____

_____

_____F~~T~~_____ 28. A permit may not be revoked permanently.

_____

_____

Multiple Choice

C. **Write the letter of the best answer on the blank before each statement.**

 1. Which of the following is the Standard for Professional Qualifications for Fire Inspector?
   a. IFSTA 1998-A   c. OSHA 14.D.12
   b. NFPA 1031   d. NFPA 101

 2. Who should provide fire inspectors with information regarding the statutes they are responsible for enforcing?
   a. The jurisdiction's governing agency attorneys
   b. The jurisdiction's legal counsel
   c. The inspectors' personal
   d. Fire department management

D 3. In the United States, who determines the legal status of most public employees?

a. Municipal government
c. State government

b. County/parish government
d. Federal government

B 4. Which of the following statements regarding insurance companies is true?

a. Insurance companies' protection requirements are rarely as strict as local codes or ordinances.

b. Some insurance company requirements are targeted more at property protection than at life safety.

c. The insurance company inspector's primary goal is to make certain that client companies enforce local codes and ordinances.

d. Insurance company inspectors can override appropriate corporate procedures when a hazard is especially dangerous.

A 5. What is the key issue when a property owner's request to modify a code involves using alternative methods of meeting the code's intent?

a. Whether the substitution will provide an equal or greater level of protection than the code requires

b. Whether the jurisdiction's authority will be compromised

c. Whether the jurisdiction can adequately inspect the substitution

d. Whether the substitution will satisfy insurance company requirements

C 6. Under most codes, to whom can an applicant appeal if he/she feels a decision is unfair?

a. District attorney

b. State fire marshal

c. Federal arbitration panel

d. Board of appeals

D 7. With which issue involving an appeal should a fire inspector be particularly familiar?

a. Cost to the jurisdiction

b. Effect on code enforcement for other properties

c. Degree of public support

d. Cost to the property owner

D 8. What is usually the first step in the enforcement procedure?

a. Giving the responsible party a spoken warning

b. Filing for an injunction

c. Notifying the responsible party in writing of the violations

d. Issuing a citation

_A_ 9. What roles does the fire inspector play in most court cases?

    a. Witness and courtroom advisor for the prosecution

    b. Defendant and witness for the defense

    c. Witness and impartial observer

    d. Media liaison and courtroom advisor for the prosecution

_A B_ 10. In cases where federal, state, and local requirements conflict, which requirement generally applies?

    a. Federal   c. Local

    b. The most lenient   d. The most stringent

_B_ 11. When does a revised edition of a model code take effect?

    a. When the code organization releases it

    b. When the authority having jurisdiction formally and legally adopts it

    c. When the courts approve it

    d. When the fire inspector accepts it

_C_ 12. Which of the following objectives is performance-based?

    a. A roof assembly must use a minimum of 2×10-inch lumber for its main structural support.

    b. A cedar shake roof must be treated with fire-retardant chemicals.

    c. A roof must be able to support a load of 40 pounds per square foot _(195 kg/m2)_.

    d. The rafters in a roof assembly must be spaced no more than 24 inches _(600 mm)_ on center.

_B A B_ 13. Which statement about performance-based objectives is **not** true?

    a. They require creative, alternate solutions.

    b. They are easier to enforce than traditional code specifications.

    c. Their administration requires significant technical expertise.

    d. Their enforcement requires more frequent inspections than traditional codes.

_D A_ 14. Which of the following is a valid statement?

    a. In recent years the federal government has chosen to comply with local codes.

    b. Local fire inspectors must ignore hazards or violations within federal properties.

    c. Local fire inspectors should avoid any official contact with the people who are responsible for federal properties.

    d. The federal government is becoming more assertive of its right to intervene in local codes.

D 15. What is the fire inspector's responsibility regarding state agencies with fire-related concerns?

    a. Maintain professional silence

    b. Enforce their statues and regulations

    c. Train them to enforce the local code

    d. Communicate to them the authority of the local jurisdiction

A 16. Which best describes the twofold purpose of the permitting process?

    a. To ensure that conditions meet code requirements and to satisfy insurance company demands

    b. To ensure that conditions meet code requirements and to keep fire department personnel aware of hazards

    c. To ensure that conditions meet code requirements and to raise funds for code enforcement

    d. To keep fire department personnel aware of hazardous conditions and to raise funds for code enforcement

B 17. What should the fire inspector do when a citizen asks if he/she needs a permit?

    a. Refer him/her to the public information office.

    b. Give permission to begin the activity.

    c. Explain the permit process.

    d. Issue the permit if the request meets applicable requirements.

B B 18. When applying for a permit, which of the following additional documentation is commonly required?

    a. Personnel sheets and budget

    b. Shop drawings and MSDS

    c. Plot diagrams and project justification report

    d. Construction drawings and proof of ownership

C A 19. Who has the authority to revoke a permit when its stipulations are not being followed?

    a. Fire inspector   c. Judge

    b. District attorney   d. Elected official

A A 20. Who typically manages a fire department's inspection activities?

    a. Fire chief

    b. Public relations director

    c. Public education officer

    d. Fire marshal

_O_ _B_ 21. Which statement below is valid?

    a. Fire prevention personnel must be sworn members of the fire department.

    b. In-service companies should not perform fire inspection activities.

    c. In some jurisdictions the fire department is not responsible for fire code enforcement.

    d. The fire marshal does not typically report directly to the fire chief.

_C_ 22. Who is usually the fire inspector in municipalities that have a separate code enforcement department?

    a. A civilian employee of the code enforcement department

    b. A fire department employee assigned to the code enforcement department

    c. A fire department employee independent of the code enforcement department

    d. A civilian code enforcement department employee assigned to the fire department

_C_ _B_ 23. Which statement below correctly describes a fire inspector's interaction with related agencies?

    a. Fire inspectors must act independently of other local agencies.

    b. When another agency's codes conflict with the fire code, the fire code usually takes precedence.

    c. The fire inspector may have to interact with the Department of Labor.

    d. The fire inspector usually is not responsible for reporting important information to other agencies.

Identify

D. **Identify the following abbreviations associated with fire codes. Write the correct interpretation before each.**

_____ 1. NFPA

_____ 2. BOCA

_____ 3. ICBO

_____ 4. SBCCI

_____ 5. MSDS

E. **Identify generally accepted legal guidelines for conducting fire inspections. Write an *X* before each correct statement below.**

    X   1.   Inspectors must be adequately identified.

           2.   Inspectors must request permission for an inspection but are not required to state the reason for the inspection.

    X   3.   Inspectors should invite a building representative to walk along during the inspection.

           4.   Electrical, plumbing, and building inspectors should not be allowed to participate in fire inspections.

           5.   Inspectors should carry and follow a written inspection procedure.

           6.   If entry is denied, inspectors should automatically issue a stop order.

           7.   All licenses and permits should indicate that compliance inspections can be made throughout the duration of the license or permit.

           8.   Inspectors should develop a reliable record-keeping system for inspections.

F. **Identify valid suggestions for courtroom procedure and behavior. Write an *X* before each correct statement below.**

           1.   The inspector should not return to the facility once a trial is scheduled.

           2.   It is illegal for fire department personnel to go over testimony with the prosecution before entering the courtroom.

           3.   Fire department personnel should appear in proper uniform or be neatly dressed.

           4.   Fire department personnel should include in their testimony any relevant information from third parties.

           5.   Fire department personnel should remain impartial and be especially careful not to indicate any dislike of the defendant.

           6.   Fire department personnel should volunteer significant information that the prosecution forgets to request.

           7.   If a fire department witness feels the need to explain an answer, he/she has the right to request the court's permission to do so.

      8. If a fire department witness does not know the answer to a question, he/she should simply say so.

      9. Fire department witnesses are allowed to refer to notes during their testimony.

      10. Fire department personnel should never become argumentative on the witness stand.

Chapter 2 Inspection Procedures

A. Match to their definitions terms associated with fire inspection procedures.

_____ 1. Indicates how a building is situated with respect to other buildings and streets in the area.

_____ 2. Shows the layout of individual floors and the roof

_____ 3. Shows the number of floors in a building and the grade around the building

_____ 4. The part of a business letter that contains the return address

_____ 5. The part of a business letter that includes the name, title, and address of the person receiving the letter

a. Floor plan

b. Inside address

c. Elevation drawing

d. Plot plan

e. Perspective drawing

f. Heading

True/False

B. Write _TRUE_ or _FALSE_ before each of the following statements. Correct those statements that are false.

_____ 1. Convincing an occupant to make a fire safety change voluntarily is easier than using legal methods to force the change.

_____

_____

_____ 2. A correctly conducted fire inspection is solely a code enforcement program.

_____

_____

_____ 3. The fire inspector should criticize the architect or builder when he finds obvious construction problems.

_____

_____

_____ 4. Accepting a soda or cup of coffee from an occupant is a good way for a fire inspector to establish a friendly rapport with an occupant during an inspection.

_____

_____

_____ 5. Fire inspectors should assume that occupants are ignorant of the fire codes.

_____

_____

_____ 6. Reviewing previous inspection records can alert an inspector to important changes in an occupancy.

_____

_____

_____ 7. The highest-ranking employee at a facility must accompany the fire inspector during a surprise inspection.

_____

_____

_____ 8. Fire inspectors may allow occupants to cover any secret business processes.

_____

_____

_____ 9. Photographs are more useful than drawings or sketches for documenting a serious problem at the time of an inspection.

_____

_____

_____ 10. Field sketches should be drawn to scale with a straightedge.

_____

_____

_____ 11. Fire inspectors should complete a business communication course.

_____

_____

_____ 12. Fire inspection personnel must maintain their files according to procedures established by the National Fire Protection Association.

_____

_____

_____ 13. Most fire departments that keep records on computers should develop their own inspection data management system.

_____

_____

Multiple Choice

C. **Write the letter of the best answer on the blank before each statement.**

_____ 1. What are the two major factors in successfully handling an emergency incident?
   a. Pre-incident planning and the occupants' cooperation during the incident
   b. The tendency of personnel to follow a pre-incident plan and occupants' cooperation during the incident
   c. Pre-incident planning and the tendency of personnel to follow the plan
   d. Pre-incident planning and the occupants' independent preparations

_____ 2. What is the key to success for fire inspectors?
   a. Communicating and dealing with people
   b. Preparing for the inspection
   c. Performing a follow-up inspection
   d. Maintaining inspection files and records

_____ 3. In what situation do most people have contact with the fire department?

    a. An emergency

    b. A fire and life safety education program

    c. A fire station tour

    d. An inspection

_____ 4. In addition to conveying the fire code's intent, what should fire inspectors be able to communicate to the public?

    a. The legislative authority for the code and the penalties for code violations

    b. Why the codes are written and historical accounts of tragic fires

    c. Why the codes are written and who writes them

    d. Historical accounts of tragic fires and the legislative authority for the code

_____ 5. What sort of attitude should a fire inspector project toward the occupant during an inspection?

    a. Helpful   c. Aggressive

    b. Fault-finding   d. Submissive

_____ 6. What should a fire inspector first be sure of before discussing a problem or violation with an occupant?

    a. A witness is present.

    b. The occupant's legal counsel is present.

    c. The fire inspector's points are correct.

    d. The occupant understands his legal obligations.

_____ 7. What information should fire inspectors convey to occupants about problems found?

    a. Suggestions for corrections and an explanation of the appeals process

    b. A citation and the amount of the fine for noncompliance

    c. A citation and an explanation of the appeals process

    d. A full explanation of the problems and suggestions for their correction

_____ 8. Which of the following factors should be considered when scheduling an appropriate amount of time for an inspection?

    a. The inspector's personal acquaintance with the occupant

    b. The insured value of the occupancy

    c. The inspector's other commitments

    d. The inspector's familiarity with the occupancy

_____ 9. In most cases, when should occupancies be inspected?

    a. During their normal business and operating hours

    b. When the fewest people are present

    c. During the inspector's normal working hours

    d. In daylight during good weather

_____ 10. When should fire inspectors begin making observations regarding the exterior of the premises?

    a. Just before entering the structure to begin an inspection

    b. After entering the premises and introducing themselves

    c. At the end of an inspection

    d. One or two days before the scheduled interior inspection

_____ 11. Where should fire inspectors enter a building when beginning an inspection?

    a. The most convenient entrance

    b. The main public entrance

    c. The least conspicuous entrance

    d. A designated service entrance

_____ 12. For whom should a fire inspector ask upon entering the premises for a surprise inspection?

    a. The owner's legal counsel

    b. Any full-time employee

    c. The manager or highest-ranking employee

    d. The person responsible for compliance with codes

_____ 13. Which of the following is a common method for performing an inspection?

    a. Start with secured areas and finish with unsecured areas.

    b. Follow an established manufacturing process from beginning to end within the structure.

    c. Start from a structure's center and work toward its perimeter.

    d. Start from the most hazardous area and work toward the least hazardous area of an occupancy.

_____ 14. What should an inspector do when an occupant refuses to unlock a locked area?

    a. Demand entry.   c. Postpone the rest of the inspection.

    b. Summon the fire chief.   d. Note it and move on.

_____ 15. Which of the following statements regarding inspection forms is true?

    a. Inspectors cannot legally look for code violations that are not included on an official inspection form.

    b. Relying on a checklist or inspection form is unprofessional.

    c. Fire departments must develop their own inspection forms.

    d. Individual inspectors may develop checklists for their own use.

_____ 16. When are formal inspection reports required?

    a. Only for major inspections or serious code violations

    b. For all inspections

    c. For all inspections that reveal one or more code violations

    d. Only for follow-up inspections involving unresolved code violations

_____ 17. What is one purpose of the closing interview?

    a. To discuss violations and the appeals process in specific terms

    b. To note good conditions as well as those that need correcting

    c. To argue specific points with hostile individuals

    d. To give the occupant a copy of the written report

_____ 18. Which of the following statements about written reports is true?

    a. Written reports should firmly express the inspector's opinions.

    b. Written evidence of an inspection is the only proof that the fire inspector gave the owner notice of hazardous conditions.

    c. Written reports do not have to include the name of the person who accompanied the fire inspector during the inspection.

    d. Written reports should not be used to draw conclusions or justify recommendations.

_____ 19. Which of the following statements is most appropriate for a business letter?

    a. "Your communication of June 29, 1998, is in hand."

    b. "This is to acknowledge receipt of your letter."

    c. "Thank you for your letter."

    d. "Got your letter, thanks!"

_____ 20. Which of the following statements about business letters is true?

    a. The reader will interpret the letter's appearance as an example of the inspector's professionalism.

    b. A letter's style is more important than its appearance.

    c. Inspectors should adapt the department's standard format to reflect their own taste.

    d. "7/6/98" is an acceptable date line.

____ 21. What is the common practice for salutations when the writer does not know the name of the person who will receive the letter?

    a. "Dear Sir or Madam"

    b. "Gentlemen"

    c. "To whom it may concern"

    d. A subject line

____ 22. What is an appropriate action for a follow-up inspection?

    a. Issue citations to occupants who have made no effort to correct problems.

    b. Revoke permits and certificates if problems remain.

    c. Compliment occupants who have corrected some problems but must still correct others.

    d. Reinspect the entire occupancy.

____ 23. What is generally the most reliable method of cataloging inspection records?

    a. Alphabetically by occupant name

    b. Chronologically by date of last inspection

    c. By the building's street address

    d. By the owner's social security number

Identify

D. Identify minimum information that an inspection letter or report should include. Write an *X* before each correct statement below.

____ 1. Name of business and type of occupancy

____ 2. Inspector's professional qualifications

____ 3. Business owner/occupant's federal employer identification number (FEIN)

____ 4. Property owner's social security number

____ 5. Name of person accompanying inspector during inspection

____ 6. Edition of applicable code

____ 7. Inspector's assessment of occupant's attitude

____ 8. Recommendations for correcting all code violations

E. **Identify minimum requirements for maintaining inspection files within a jurisdiction. Write an *X* before each correct statement below.**

_____ 1. The jurisdiction should maintain files on all one- and two-family residences.

_____ 2. The jurisdiction should maintain files on all occupancies that have been issued a permit, certificate, or license.

_____ 3. The jurisdiction should maintain files on occupancies that contain automatic fire suppression or detection systems.

_____ 4. The jurisdiction should maintain all records indefinitely.

_____ 5. The fire prevention bureau's records on privately owned occupancies are considered confidential.

_____ 6. The jurisdiction should maintain files on all occupancies that routinely contain hazardous materials.

Chapter 3   Principles of Combustion and Fire Growth

A. **Match to their definitions terms associated with principles of combustion and fire growth. Definitions are continued on the next page.**

_____ 1. The self-sustaining process of rapid oxidation of a fuel

_____ 2. The result of a rapid combustion reaction

_____ 3. Very slow oxidation

_____ 4. Illustrates the smoldering mode of combustion

_____ 5. Illustrates the flaming mode of combustion

_____ 6. "Matter in motion" caused by the movement of molecules

_____ 7. The amount of heat generated by oxidation

_____ 8. The heating of an organic substance without the addition of external heat

_____ 9. Heat generated by passing an electrical current through a conductor

_____ 10. Heating that occurs when electrical current flow is interrupted

_____ 11. Energy generated when atoms split apart

_____ 12. Any solid, liquid, or gas that can combine with oxygen in the chemical reaction known as oxidation

_____ 13. Takes place when unburned combustible gases accumulate at the ceiling level

a. Heat from arcing

b. Heat

c. Fire load

d. The fire tetrahedron

e. Impregnation

f. Draft curtains

g. Fuel

h. Resistance heating

i. Fission

j. The fire triangle

k. Spontaneous heating

l. Fire

m. Convection

n. Rusting

o. Fire-stop systems

p. Fusion

q. Heat of combustion

r. Combustion

s. Fire-stops

t. Rollover

_____ 14. The most common method of heat and fire travel within structures

_____ 15. The maximum heat that can be produced if all the combustibles in a given area burn

_____ 16. Structural element designed to limit the mushrooming effect of heat and smoke in large open areas of buildings

_____ 17. Pieces of 2-inch by 4-inch lumber placed between wall studs in wood-frame construction

_____ 18. Constructed as needed to close utility penetrations in fire walls

_____ 19. Technique in which an absorbent material is saturated with a fire-retardant agent

True/False

B. **Write** _TRUE_ **or** _FALSE_ **before each of the following statements. Correct those statements that are false.**

_____ 1. When air is not the oxidizing agent of a fire, chlorine or chemicals that release chlorine must be.

_____

_____

_____ 2. Leakage current heating occurs when a wire is not insulated well enough to contain all the current.

_____

_____

_____ 3. A positive heat imbalance occurs when heat is dissipated faster than it is generated.

_____

_____

_____ 4. According to the Law of Heat Flow, the colder of two bodies in contact will absorb heat until both objects are at the same temperature.

_____

_____

_____ 5. Convection is the transfer of heat by the movement of air or liquid.

_____

_____

_____ 6. Direct flame contact is a form of convective heat transfer.

_____

_____

_____ 7. Fire resistance is a structure's ability to prevent combustion.

_____

_____

_____ 8. The fire-resistive rating represents the period of time an assembly will perform satisfactorily when exposed to a standard test fire.

_____

_____

_____ 9. Hose stream tests are only used for fire doors.

_____

_____

_____ 10. A material with a flame spread rating of 73 is less combustible than a material with a rating of 39.

_____

_____

_____ 11. Fire load ratings accurately indicate fire severity with combustibles that have a high heat of combustion.

_____

_____

_____ 12. A rain test measures how many inches *(millimeters)* of rain per hour are needed to extinguish a fully involved roof covering.

_____

_____

_____ 13. A Class B roofing material is more fire retardant than a Class C material.

_____

_____

_____ 14. A roof's fire rating is for the whole deck assembly, not just the outer covering.

_____

_____

_____ 15. A *rated fire door* is a door that has complied with the requirements of ASTM Test E512.

_____

_____

_____ 16. A fire door assembly includes only the fire door and door frame.

_____

_____

_____ 17. Fire doors will not be effective unless they are locked against egress.

_____

_____

_____ 18. Once draft curtains are in place, a fire inspector just needs to make sure that no alterations have been made that would decrease their effectiveness under fire conditions.

_____

_____

_____ 19. A disadvantage of venting a building involved in a fire is reduced visibility for occupants and firefighters.

_____

_____

_____ 20. Acceptable materials for fire-stop systems include gypsum board, mineral fiber insulation, and sand.

_____

_____

_____ 21. A flammable decoration's degree of hazard depends on how easily it can be ignited by cigarettes, sparks, electrical defects, and similar heat or ignition sources.

_____

_____

_____ 22. When applied properly to Christmas trees, boric acid is an effective fire-retardant substance.

_____

_____

Multiple Choice

C. Write the letter of the best answer on the blank before each statement.

_____ 1. The normal oxygen content in air is _____ percent.
   a. 78   c. 21
   b. 57   d. 7

_____ 2. What element normally comprises more than 75 percent of air?

    a. Hydrogen   c. Neon

    b. Oxygen   d. Nitrogen

_____ 3. What is another name for steady-state fires?

    a. Reignition fires

    b. Free-burning fires

    c. High energy-release fires

    d. Constant-rate fires

_____ 4. What element in the flaming mode of combustion is not present in the smoldering mode?

    a. Hydrogen   c. Reignition

    b. Chemical chain reaction   d. Three-way heat release

_____ 5. How do halon and halon replacement agents extinguish a fire?

    a. Interrupt the chemical chain reaction

    b. Eliminate oxygen

    c. Convert free oxygen to water

    d. Eliminate chlorides

_____ 6. Which of the following is *not* a general category of heat energy?

    a. Chemical   c. Operational

    b. Nuclear   d. Mechanical

_____ 7. What type of heat energy does a compost pile release?

    a. Combustion   c. Decomposition

    b. Spontaneous   d. Solution

_____ 8. What type of heat energy is released when an acid dissolved in water causes an explosion?

    a. Combustion

    b. Spontaneous

    c. Decomposition

    d. Solution

_____ 9. What type of electrical heat do microwave ovens use?

    a. Resistance   c. Friction

    b. Dielectric   d. Static

_____ 10. What type of electrical heat occurs when flammable liquids are transferred between two containers that are not properly electrically bonded together?

    a. Resistance   c. Friction

    b. Dielectric   d. Static

_____ 11. What kind of heat energy does a loose belt on a pulley generate?

    a. Friction   c. Compression

    b. Static electricity   d. Reciprocation

_____ 12. What kind of heat energy powers a diesel engine?

    a. Friction   c. Compression

    b. Static electricity   d. Reciprocation

_____ 13. When radiant heat provides energy for continued vaporization, it is called ___.

    a. Fusion   c. Radiative feedback

    b. Steam perpetuation   d. Thermonuclear radiation

_____ 14. The earliest phase of a fire is called the ___ phase.

    a. Ignition   c. Postignition

    b. Incipient   d. Nascent

_____ 15. The phase of a fire when fire growth and open burning make total involvement possible is generally considered the ___ burning phase.

    a. Steady-state   c. Inferno

    b. Preconflagration   d. Absolute

_____ 16. What may occur when all the contents of a room reach their ignition temperatures?

    a. Rollover   c. Flashover

    b. Backdraft   d. Burnout

_____ 17. What happens during the hot-smoldering phase of a fire?

    a. The room clears of smoke.

    b. The pressure from released gases decreases.

    c. Inverse rollover is possible.

    d. Burning is reduced to glowing embers.

_____ 18. What occurs when oxygen rushes into a confined space during the hot-smoldering phase of a fire?

    a. Likely rollover   c. Possible flashover

    b. Risk of backdraft   d. Burnout

_____ 19. What kind of heat transfer occurs when a fire heats pipes enough to ignite the walls of rooms in another area of the structure?

    a. Convection   c. Transportation

    b. Radiation   d. Conduction

_____ 20. What kind of heat transfer occurs when heated gases spread a fire through an elevator shaft?

    a. Convection   c. Transportation

    b. Radiation   d. Conduction

_____ 21. What kind of heat transfer occurs when heat waves spread a fire from one structure to another?

    a. Convection   c. Transportation

    b. Radiation   d. Conduction

_____ 22. On what basis do building codes classify types of building construction?

    a. Square-foot _(square meter)_ ratings

    b. Assessed valuation

    c. Fire-resistance ratings

    d. Occupancy use

_____ 23. The Steiner Tunnel Test's numerical expression of a material's surface combustibility is the ___.

    a. Flame spread rating

    b. Fire-resistance rating

    c. Ignitability rating

    d. Flammability index

_____ 24. Which of the following is **not** another designation for the Steiner Tunnel Test?

    a. UL 723   c. NIST 75.32

    b. ASTM E84   d. NFPA 255

_____ 25. What reference gives the flame spread ratings for construction materials?

    a. _Contractor's Fire Guide_

    b. _Construction Materials Manual_

    c. _Builder's Code Index_

    d. _Building Materials Directory_

_____ 26. What does the smoke-developed rating measure?

    a. Toxicity of the products of combustion

    b. The visual obscurity created by smoke generated by a specific material

    c. Parts per million of carbon in the smoke of a specific material

    d. The size of the smoke particles relative to those in smoke from burning red oak

_____ 27. Typically, what would be the fire load of a welding area containing few combustibles?

    a. Slight   c. Moderately severe

    b. Moderate   d. Severe

_____ 28. Typically, what would be the fire load of a machine shop with noncombustible floors?

    a. Slight   c. Moderately severe

    b. Moderate   d. Severe

_____ 29. How is a typical fire load greater than 20 psf _(98 kg/m²)_ classified?

    a. Slight   c. Moderately severe

    b. Moderate   d. Severe

_____ 30. What current development will eventually make traditional fire load ratings obsolete?

    a. Computerized fire modeling programs

    b. Fireproof building materials

    c. Formation of a national fire load rating committee

    d. Better fire detection and suppression systems

_____ 31. Which roof covering test determines whether burning brands are likely to ignite a roof covering?

    a. Flying brand test

    b. Airborne ember test

    c. Windblown object conduction test

    d. Burning brand test

_____ 32. What structural device prevents flames on a roof from spreading beyond a fire wall?

    a. Rampart   c. Cupola

       Parapet   d. Gutter

3

_____ 33. Which of the following characteristics is required for fire walls?

    a. No penetrations for utility extension

    b. Self-supporting

    c. No doors

    d. Properly protected penetrations with combustible materials

_____ 34. What distinguishes a fire partition from a fire wall?

    a. A fire partition has a higher fire resistance.

    b. A fire partition may not have any doors.

    c. A fire partition must extend through and above any combustible roof.

    d. A fire partition does not have to be self-supporting.

_____ 35. Which type of activation device must be used in fire doors designed for life safety purposes?

    a. Fusible link

    b. Rolling pin

    c. Electronic

    d. Water pressure

_____ 36. Where are counterbalanced doors generally used?

    a. Openings to freight elevators

    b. Openings in fire walls

    c. Openings for emergency exit

    d. Openings in floors and roofs

_____ 37. What characteristic should a fire inspector look for when inspecting fire doors?

    a. A higher rating than the fire wall

    b. Painted fusible links and rollers

    c. No windows in fire doors used for stairwells

    d. A clear path for easy opening and closing

_____ 38. In general, where would a fire inspector _least_ expect to find smoke and heat vents?

    a. Large, single-story structures

    b. Windowless structures

    c. Structures with automatic sprinkler systems

    d. Underground structures

_____ 39. What three types of smoke and fire vents are commonly used today?

    a. Manually operated vents, automatic unit vents, and mechanical venting systems

    b. Monitors, sawtooth skylights, and continuous gravity vents

    c. Breakable glass, stationery shutters, and automatic unit vents

    d. Manually operated vents, sawtooth skylights, and mechanical venting systems

_____ 40. Which of the following statements about fire and smoke dampers is true?

    a. Fire dampers manually interrupt the air flow through all or part of an air handling system.

    b. Smoke dampers are required in air ducts that pass through smoke barrier partitions.

    c. Fire dampers are usually designed for horizontal installation.

    d. Most fire dampers have a heat-resistive rating of at least 3 hours.

_____ 41. What does a fire-stop's _T_ rating measure?

    a. The temperature at which the fire-stop will ignite

    b. Whether combustible materials on the side not exposed to fire will ignite through conduction

    c. The fire-stop's ability to maintain its physical integrity when exposed to fire and hose streams

    d. The minimum amount of time that the fire-stop will prevent fire spread

_____ 42. Which method of making a material fire retardant may be done at any time during the material's life?

    a. Chemical change   c. Pressure impregnation

    b. Impregnation   d. Coating

_____ 43. Which type of fire retardant swells to a puffy form that provides a heat insulating barrier and excludes oxygen from the fuel?

    a. Mastics

    b. Gas-forming paints

    c. Intumescent paints

    d. Cementitious and mineral fiber coatings

_____ 44. Which type of fire retardant has been used primarily on steel structural members in the past?

    a. Mastics

    b. Gas-forming paints

    c. Intumescent paints

    d. Cementitious and mineral fiber coatings

_____ 45. To field test a material's flammability, the fire inspector should hold a flame tip
_____.

    a. $^1/_4$ inch *(6 mm)* below the material for 5 seconds

    b. $^1/_2$ inch *(13 mm)* below the material for 5 seconds

    c. $^1/_2$ inch *(13 mm)* below the material for 12 seconds

    d. 1 inch *(25 mm)* below the material for 12 seconds

_____ 46. During a field test for flammability, the material should not support combustion or
continue to flame for more than _____ seconds after the test flame is removed.

    a. 2 seconds

    b. 4 seconds

    c. 6 seconds

    d. 8 seconds

Identify

D. **Identify the following abbreviations associated with fire-resistance testing.
Write the correct interpretation before each.**

_____ 1. UL

_____ 2. NIST

_____ 3. FM

_____ 4. ASTM

_____ 5. NBS

Chapter 4 Fire Hazard Recognition

A. Match to their definitions terms associated with electrical fire hazards.

_____ 1. The transfer or movement of electrons between atoms

_____ 2. Material that allows the free movement of large numbers of electrons

_____ 3. Material that is neither a good conductor nor a good insulator

_____ 4. Quantity of electricity

_____ 5. Electrical pressure

_____ 6. Resistance in an electrical circuit

_____ 7. A nonflowing electrical charge

_____ 8. Connecting two objects that conduct electricity with another conductor

_____ 9. Connecting an object that conducts electricity to the ground with another conductor

a. Amperes

b. Static

c. Grounding

d. Electricity

e. Static discharge

f. Ohms

g. Bonding

h. Voltage

i. Semiconductor

j. Conductor

B. Match to their definitions terms associated with HVAC systems, temporary/portable heating equipment, and cooking equipment. Terms and definitions are continued on next page.

_____ 1. Furnace in which combustion gases pass through tubes immersed in circulating water

_____ 2. Furnace in which combustion gases pass over drums and tubes containing steam and water

a. Salamander

b. Two-position nozzle

c. Primary safety control

_____ 3. Unit that relies on a fan to move heated air through the system

_____ 4. Self-contained, automatically controlled heating appliance

_____ 5. Unit that uses radiant heat to warm an area

_____ 6. Device that stops fuel flow to a unit in the event of an ignition or flame failure

_____ 7. Substance that is toxic in concentrations of less than 400 ppm by volume

_____ 8. Small LPG-, propane-, or solid-fuel heater common at construction sites

_____ 9. Fire protection system especially calculated and constructed for a specific hazard

_____ 10. Discharge device that shoots a straight stream of agent

_____ 11. Discharge device that projects a fan-shaped fine stream of agent

d. Water-tube boiler

e. Room heater

f. Engineered system

g. Class A refrigerant

h. Class B refrigerant

i. One-position nozzle

j. Forced air furnace

k. Fire-tube boiler

l. Unit heater

C. **Match to their definitions terms associated with flammable finishes, dip tanks, and quenching operations.**

_____ 1. Coating process that usually uses an air spray gun

_____ 2. Coating process that suspends electrostatically charged particles in air

_____ 3. Device in which moving racks pass parts through a fluid- or powder-coating enclosure

_____ 4. Device used to apply paint strips or other patterns on automobile parts

a. Decorating machine

b. Dip tank

c. Dry filter

d. Freeboard

e. Fluid coating

f. Cascade scrubber

_____ 5. Device in which reusable panels accumulate particles from paint overspray

_____ 6. Device that captures paint overspray particles in water

_____ 7. Device in which parts are immersed in liquid

_____ 8. The distance from the liquid surface to the top of a fully-loaded quenching tank

g. Baffle maze

h. Powder coating

i. Continuous coater

D. Match to their definitions classes of furnaces.

_____ 1. Operate at approximately atmospheric pressure; present potential for explosion or fire hazard when heating flammable volatiles or combustible materials

_____ 2. Operate at temperatures ranging from above ambient to over 5,000°F *(2 760°C)* at pressures normally below atmospheric

_____ 3. Operate at approximately atmospheric pressure; heat no flammable volatiles or

a. Class A

b. Class B

c. Class C

d. Class D

True/False

E. Write *TRUE* or *FALSE* before each of the following statements. Correct those statements that are false.

_____ 1. Most electrical fires are caused by static discharge.

_____

_____

_____ 2. Currents of less than 200 milliamperes cannot cause death.

_____

_____

_____ 3. Maintaining a relative humidity of 60 to 70 percent greatly reduces the static electricity problems associated with manufacturing paper, cloth, and fiber.

_____

_____

_____ 4. Driveways between lumber stacks in an open-storage lumber yard should be a minimum of 20 feet *(6 m)* wide.

_____

_____

_____ 5. Incineration is effective in destroying medical, chemical, and biological wastes.

_____

_____

_____ 6. Making sure that incompatible products are kept away from each other in warehouses is a simple matter in most jurisdictions.

_____

_____

_____ 7. During warehouse inspections, fire inspectors should gather information about plant personnel's level of training.

_____

_____

_____ 8. Shop rags or towels stored in welding or cutting areas should be kept in approved metal containers with lids.

_____

_____

_____ 9. A hot work program should include a form that requires the person requesting the permit to ensure adherence to all fire prevention safeguards.

_____

_____

_____ 10. The mechanical equipment in an HVAC system must be enclosed in a separate room with a minimum fire-resistance rating of one hour.

_____

_____

_____ 11. Class 2 HVAC filters, when clean, will not contribute to a fire and will emit very small quantities of smoke when attacked by flames.

_____

_____

_____ 12. An active smoke control system may exhaust combustion products through the building's HVAC system.

_____

_____

_____ 13. Some types of heaters may be illegal in specific types of occupancies.

_____

_____

_____ 14. The containers used in a dry-chemical fire protection system are basically the same as those used in portable fire extinguishers.

_____

_____

_____ 15. Wet chemical fire protection systems are recommended for electrical fires.

_____

_____

_____ 16. The finishing area of an occupancy includes the interiors of the spray area's exhaust ducts.

_____

_____

_____ 17. Buildup of paint materials on walls of a spray booth can act as an insulator and render grounding equipment useless.

_____

_____

_____ 18. Dipping processes should be located in an egress area whenever possible.

_____

_____

_____ 19. Any oil quenching tank over 150 gallons _(600 L)_ must have a bottom drain that will open automatically or manually in the event of a fire.

_____

_____

_____ 20. The first in a series of dust explosions is usually the most severe.

_____

_____

_____ 21. The dust hogger is the most common location of fires at woodworking operations.

_____

_____

_____ 22. Metal dusts present a considerably lower explosion hazard than grain dust or wood dust.

_____

_____

_____ 23. A fire inspector must actually visit a site proposed for open burning before issuing a permit.

_____

_____

_____ 24. Fires in approved burn containers should not be within 50 feet *(15 m)* of a structure.

_____

_____

_____ 25. Inside storage of tires should be in accordance with local code requirements or NFPA 231D.

_____

_____

_____ 26. HPMs may be stored in specially designed rooms within the manufacturing building itself.

_____

_____

Multiple Choice

F. **Write the letter of the best answer on the blank before each statement.**

_____ 1. Which of the following materials is not a good conductor?
   a. Copper   c. Steel
   b. Aluminum   d. Glass

_____ 2. What does *I* stand for in Ohm's Law *(E = I · R)?*
   a. Pressure   c. Electricity
   b. Current   d. Resistance

_____ 3. Which of the following characteristics of electricity corresponds to friction loss in a fire hose?

    a. Amperage

    b. Static

    c. Ohms

    d. Voltage

_____ 4. Why is battery recharging especially hazardous?

    a. Poisonous carbon monoxide can be produced during this operation.

    b. Too high a voltage can cause excessive pressure in the container.

    c. Corrosive gases may damage the lungs.

    d. Hydrogen gas produced during this operation can ignite explosively.

_____ 5. When can static discharge occur?

    a. When there is a good electrical path between two materials

    b. At a minimum of 120 volts AC

    c. When there is no good electrical path between two materials

    d. When two materials are grounded and bonded

_____ 6. Which means of ionizing the air to dissipate static charges is applicable to processes involving cotton, wool, or silk?

    a. High voltage device

    b. Open flame

    c. Glass wand

    d. Static comb

_____ 7. For open yard storage, what is the maximum recommended height for stacks of wood in a lumberyard?

    a. 10 feet _(3 m)_

    b. 15 feet _(4.5 m)_

    c. 20 feet _(6 m)_

    d. 25 feet _(7.5 m)_

_____ 8. What are the maximum recommended dimensions for a driveway grid in an open-storage lumberyard?

    a. 50 feet by 150 feet _(15 m by 45 m)_

    b. 100 feet by 150 feet _(30 m by 45 m)_

    c. 50 feet by 100 feet _(15 m by 30 m)_

    d. 100 feet by 200 feet _(30 m by 60 m)_

_____ 9. What resource should a fire inspector consult for specific requirements for lumberyards?

    a. *The National Lumberman's Fire Protection Guide*

    b. NFPA 505

    c. *Underwriter's Manual*

    d. Local codes

_____ 10. What are the two main hazards associated with recycling plants?

    a. High occupancy and bulk storage of hazardous materials

    b. Hazardous processes and bulk storage of combustible materials

    c. Bulk storage of combustible materials and hazardous materials

    d. Hazardous processes and high occupancy

_____ 11. What is the recommended minimum distance between sprinkler heads and the tops of material stacks in a recycling plant?

    a. 1 foot *(0.3 m)*

    b. 18 inches *(450 mm)*

    c. 2 feet *(0.6 m)*

    d. 30 inches *(750 mm)*

_____ 12. How should recycling plants store hot ashes, coals, cinders, or material subject to spontaneous heating?

    a. Off premises in glass containers

    b. In cement vaults at least 15 feet *(4.5 m)* from stored combustibles

    c. In approved metal containers at least 10 feet *(3 m)* from stored combustibles

    d. In slurry vats

_____ 13. How much can incineration reduce the bulk of waste?

    a. 75 percent

    b. 90 percent

    c. 95 percent

    d. 99 percent

_____ 14. Which statement about waste disposal operations is true?

    a. Dense smoke and serious odors should be vented to the outside.

    b. Fuel-fire incinerators should not be preheated.

    c. Feed doors may remain open only 2 minutes after the burning cycle begins and must remain fully closed for the remainder of the combustion cycle.

    d. Fuel-fire incinerators should be preheated for 30 minutes.

_____ 15. What is one way to reduce hazards at a waste disposal facility?

    a. Storing all combustible materials in one building

    b. Allowing waste to stabilize 72 hours before disposing of it

    c. Marking containers and areas and ensuring their integrity

    d. Offsite decontamination and sanitation facilities

_____ 16. What storage method consists of a structural framework onto which pallets or other materials are placed?

    a. Rack   c. Post and Beam

    b. Palletized   d. Solid piling

_____ 17. Which storage method gives fires the least chance of developing?

    a. Rack   c. Post and Beam

    b. Palletized   d. Solid piling

_____ 18. When is the best time to inspect a warehouse?

    a. When it is near capacity

    b. When relatively fewer materials are present

    c. When it is relatively less busy

    d. When employees are changing shifts

_____ 19. What is most critical for an inspector to check at a warehouse?

    a. Properly posted OSHA regulations

    b. Clearly marked exits

    c. Automatic sprinkler system

    d. Licenses and permits

_____ 20. Which two elements of the fire triangle are inherent in welding and thermal cutting?

    a. Oxygen and fuel

    b. Heat and oxygen

    c. Heat and fuel

    d. Self-sustained chemical reaction and heat

_____ 21. Which of the following gases is most commonly used for oxy-fuel gas welding?

    a. Propane

    b. Acetylene

    c. Methane

    d. Compressed natural gas

_____ 22. What are the two primary fire safety concerns with welding and thermal cutting?

    a. Adequate distance from combustible materials and licensed operators

    b. Properly maintained fire detection system and safe equipment

    c. Licensed operators and safe equipment

    d. Safe equipment and adequate distance from combustible materials

_____ 23. The primary fire hazard associated with oxy-fuel gas equipment is ___.

    a. Properly maintained cutting torch

    b. Corroded fuel lines

    c. Storage of the oxygen and fuel gas cylinders

    d. Improperly trained operators

_____ 24. How large an area should be protected from sparks or slag at welding or cutting operations?

    a. 25 feet *(7.5 m)*   c. 35 feet *(11 m)*

    b. 30 feet *(9 m)*   d. 45 feet *(14 m)*

_____ 25. When a fire watch is required at welding or thermal cutting operations, how long must it be maintained?

    a. 20 minutes   c. 45 minutes

    b. 30 minutes   d. 60 minutes

_____ 26. The biggest fire safety concern related to HVAC systems is ___.

    a. Fire hazards caused by the heating appliance

    b. Combustible gases in the refrigeration system

    c. Spread of fire products through the air handling system

    d. Sparks from the AC compressor

_____ 27. What is the primary fire hazard associated with boilers?

    a. Thermal leakage

    b. Contact with combustible materials

    c. High-velocity convection

    d. Explosion

_____ 28. Where are active smoke control systems especially applicable?

    a. Warehouses

    b. Single-family dwellings

    c. High-rise structures

    d. Lumberyards

_____ 29. Which of the following questions is *not* generally important when a fire inspector examines an HVAC system?

    a. Is the designer qualified and licensed?

    b. Is the air humidified/dehumidified?

    c. What type of filter coatings/adhesives are used?

    d. Where are the fire suppression systems/drains?

_____ 30. Why is proper venting of kerosene heaters imperative?

    a. To prevent buildup of explosive gases

    b. To prevent a dangerous vacuum

    c. To prevent buildup of carbon monoxide

    d. To prevent atmospheric fuel leakage

_____ 31. Which statement about portable heaters is true?

    a. A properly grounded and bonded extension cord may be used to plug an electric heater into an otherwise overloaded electrical circuit.

    b. Heaters should be equipped with safety switches that turn off the heating element if the heater overturns.

    c. Properly vented and isolated kerosene heaters may be filled within a structure.

    d. Fuels stored within a structure should be no more than 6 feet *(2 m)* from a door or window.

_____ 32. Unless specially designed for a lower clearance, cooking equipment must have a minimum clearance of ___ from any combustible material.

    a. 12 inches *(300 mm)*   c. 24 inches *(600 mm)*

    b. 18 inches *(450 mm)*   d. 30 inches *(750 mm)*

_____ 33. What standard states the requirements for the hood, exhaust, and fire protection systems above cooking areas in cooking establishments?

    a. NFPA 96

    b. OSHA 12.A

    c. NIST 1555

    d. NFPA 1031

_____ 34. Historically, how are most dry-chemical fire protection systems in cooking areas activated?

    a. Electronic sensor

    b. Frangible bulb

    c. Bimetallic strip

    d. Fusible link

_____ 35. What type of fire protection is required for solid-fuel burning systems that have a fuel box greater than 5 cubic feet *(0.14 m³)?*

    a. Dry chemical system

    b. Portable fire extinguisher

    c. Fixed water system

    d. Halon system

_____ 36. How much flammable liquid should be kept on hand at a finishing area?

    a. Only enough for one day's work

    b. Only enough for all scheduled jobs

    c. Only enough for work in progress

    d. Only enough for one shift's work

_____ 37. What is the major cause of fires in powder coating processes?

    a. Spontaneous combustion   c. Open flame

    b. Static accumulation   d. Exposed electrical wires

_____ 38. What is the normal flash point of mediums used for heated quenching?

    a. 500°F *(260°C)*   c. 300°F *(149°C)*

    b. 400°F *(205°C)*   d. 200°F *(93°C)*

_____ 39. Which of the following statements about quenching tanks is correct?

    a. Tanks should be built in basements.

    b. Tanks should be installed a minimum of 5 feet *(1.5 m)* above grade.

    c. Freeboard should never be less than 9 inches *(225 mm)*.

    d. Tanks should be built within dikes.

_____ 40. At quenching operations, how many flammable liquid storage cabinets may be allowed in a single process area without approval of the authority having jurisdiction?

    a. Two   c. Four

    b. Three   d. Five

_____ 41. Dry cleaning that also includes standard laundering either before or after the dry cleaning is ___.

    a. Added value dry cleaning

    b. Wet/dry processing

    c. Double-wash cleaning

    d. Dual-phase processing

_____ 42. What standard specifies safety requirements for dry-cleaning establishments?

    a. OSHA 178.c.2

    b. DC&L Reg. 3

    c. SASE 20109

    d. NFPA 32

_____ 43. Which of the following statements about dry-cleaning operations is correct?

    a. The roof above a dry-cleaning operation should have a fire-resistance rating of not less than 2 hours.

    b. A dry-cleaning room may have only one door if that door leads directly to the outside.

    c. Curbs or scuppers may not be used in a dry-cleaning operation's emergency drainage system.

    d. Dry-cleaning operations using combustible liquids should not be performed in the same building with other occupancies.

_____ 44. Which of the following statements regarding grain storage operations is correct?

    a. Mechanical dust control systems can collect no more than 90 percent of dust.

    b. Buildings used in grain handling should comply with local building codes and NFPA 65.

    c. Access doors must be provided to permit inspection, cleaning, and maintenance.

    d. Subterranean tunnels and passageways 50 feet _(15 m)_ or longer must have two means of egress as close to each other as possible.

_____ 45. Which of the following statements regarding grain storage operations is correct?

    a. Each facility's emergency action plan should include the location of material safety data sheets (MSDS).

    b. The NFPA recommends wet or dry standpipes be provided to all operating levels of 50 feet _(15 m)_ or higher.

    c. Fire protection for grain elevators is not comparable to fire protection for any other industry.

    d. Sacks or other supplies must be stored in areas where the only other combustible material is the agricultural commodity being stored.

_____ 46. What is the fire inspector's primary concern at woodworking facilities?

    a. Dust fires

    b. Wood finishing processes

    c. Dust explosions

    d. Machinery fires

_____ 47. Small pieces of metal that wear off woodworking machinery during milling are called ___.

    a. Hot shots   c. Fireflies

    b. Butts   d. Tramp

_____ 48. How are metal pieces removed from wood dust on conveyors?

    a. Gravity screen   c. Wind tunnel

    b. Water slurry   d. Magnetism

_____ 49. What is the most common cause of fires and explosions in industrial furnaces and ovens?

    a. Poor equipment design

    b. Human error

    c. Equipment malfunction

    d. Overheated bearings

_____ 50. Which of the following statements regarding hazards associated with ovens and furnaces is true?

    a. If a malfunction occurs, the affected area should shut down while the remaining components continue to function.

    b. Safety devices should automatically return the system to normal operation as soon as a malfunction is corrected.

    c. Ovens processing sufficient combustion materials to sustain a fire must be equipped with automatic sprinklers or water spray.

    d. Overheating and other malfunctions must be manually controlled.

_____ 51. Which of the following is not typically included in regulations that control open burning?

    a. Trash in barrels

    b. Fires used to cook food for human consumption

    c. Campfires

    d. Prescribed burns of agricultural land

_____ 52. Which of the following statements is correct?

    a. Tar kettles may be operated on the roof of a building but not inside a building.

    b. Tar kettles must be attended by at least two persons.

    c. Two approved A:B fire extinguishers shall be maintained within 25 feet _(7.5 m)_ of an operating tar kettle.

    d. Tar kettles should be no closer than 10 feet _(3 m)_ to exits or means of egress.

_____ 53. Which of the following is *not* a special concern regarding fires where tires are stored?

     a. Explosion potential   c. Toxic oil

     b. Flame intensity   d. Massive smoke

_____ 54. Which of the following statements regarding tire storage operations is *not* true?

     a. Weeds and vegetation must be cleared within 50 feet *(15 m)* of any outside tire pile.

     b. Fire codes typically specify the maximum height, length, and width of tire piles.

     c. High-expansion foam is not effective for suppressing inside tire storage fires.

     d. Fire codes typically specify the distances between a tire pile and other tire piles, property lines, and structures.

_____ 55. Which of the following statements concerning powered industrial trucks is true?

     a. The battery-charging area should be vented to prevent an accumulation of hydrogen gas.

     b. The most common types are LPG and CNG.

     c. Liquid- or gas-fueled vehicles should be fueled in a separate, restricted area inside the building.

     d. Chargers for electric battery-operated units should be at least 10 feet *(3 m)* from combustible materials.

_____ 56. Which of the following statements about semiconductor and electronics manufacturing is true?

     a. Semiconductors are inherently more hazardous than the processes used to manufacture them.

     b. Areas of particular concern at a semiconductor manufacturing operation are clean rooms and semiconductor storage areas.

     c. Dry chemical extinguishers are the best choice for extinguishing fires in clean rooms.

     d. Semiconductor clean rooms must be made of fire-resistive construction and contain a sprinkler system.

Identify

G. **Identify the following abbreviations associated with fire codes. Write the correct interpretation before each.**

_____ 1. OFW

_____ 2. HVAC

_____ 3. LPG

_____ 4. CNG

_____ 5. HPM

H.   **Identify conditions necessary for a dust explosion. Write an _X_ before each necessary condition below.**

_____ 1. Spontaneous combustion

_____ 2. Combustible dust suspended in air

_____ 3. Dust must exceed its explosive range

_____ 4. Self-sustaining chemical reaction

_____ 5. Presence of an ignition source

_____ 6. Dust in a confined space

I.   **Identify facts about construction and demolition hazards. Write an _X_ before each correct statement below.**

_____ 1. The risk of fire is lower at construction and demolition sites than at completed occupancies.

_____ 2. Contractors bring additional fire loading and ignition sources to construction and demolition sites.

_____ 3. The decreased fire loading at unfinished wood frame structures slows fire spread.

_____ 4. The lack of doors on unfinished structure contributes to rapid fire growth.

_____ 5. Buildings that are being renovated or demolished have the potential for more sudden building collapse than completed structures.

_____ 6. The risk of arson at construction or demolition sites is lower than at completed structures.

_____ 7. The property owner must appoint a fire protection manager who will develop pre-incident plans.

_____ 8. Plans, keys, and emergency information must be given to the fire department for off-site storage.

_____ 9. The last piece of fire equipment removed from a building being demolished should be the standpipe.

_____ 10. Renovated structures can be especially hazardous because they are not required to meet current codes.

_____ 11. Occupants and their belongings should not be allowed to remain in a building during renovations.

_____ 12. Trash and rubbish at construction sites should be picked up and disposed of weekly.

J. **Identify requirements concerning fire safety and tents. Write an *X* before each correct statement below.**

_____ 1. Approved flame-resistant materials must be used for any tent.

_____ 2. Banners or decorations on the outside of a tent are exempt from fire-retardant requirements.

_____ 3. The inspector should test the flammability of any tent that is not properly tagged.

_____ 4. Tents should not cover more than 60 percent of the premises.

_____ 5. Pyrotechnics and flames are prohibited in any tents or temporary membrane structures.

_____ 6. Railings or guards should not be more than 3 feet *(1 m)* above the aisle surface or footrest.

_____ 7. Hay, straw, or shavings may be allowed in the tent if they have been treated with fire-retardant chemicals.

Chapter 5   Construction and Occupancy Classifications

A.   **Match to their definitions terms associated with construction and occupancy classifications.**

_____   1.   How a building is constructed

_____   2.   What a building is used for

_____   3.   NFPA Type I, BOCA Type 1, and UBC Type I construction

_____   4.   NFPA Type II, BOCA Type 2, and UBC Type II construction

_____   5.   NFPA Type III, BOCA Type 3, and UBC Type III construction

_____   6.   NFPA Type IV, BOCA Type 4, and UBC Type IV construction

_____   7.   NFPA Type V, BOCA Type 5, and UBC Type V construction

_____   8.   Structural members are constructed, chemically treated, covered, or protected so that the entire assembly has an appropriate fire-resistance rating.

a.   Noncombustible construction

b.   Protected construction

c.   Standard construction

d.   Occupancy classification

e.   Heavy timber construction

f.   Construction classification

g.   Fire-resistive construction

h.   Ordinary construction

i.   Wood-frame construction

5

B. Write *TRUE* or *FALSE* before each of the following statements. Correct those statements that are false.

_____ 1. The only major difference between NFPA Type I and NFPA Type II construction ratings is that the degree of fire resistance is lower in Type I.

_____

_____

_____ 2. The primary difference between BOCA Type 1 and BOCA Type 2 construction classifications is in the fire-resistance ratings of various construction components.

_____

_____

_____ 3. The primary difference between UBC Type III-One-Hour and UBC Type III-N is that Type III-One-Hour requires one-hour fire-resistive construction throughout the entire structure.

_____

_____

_____ 4. UBC Type IV buildings may have temporary partitions.

_____

_____

_____ 5. SBC construction Types I and II are similar to BOCA Types 1 and 2.

_____

_____

_____ 6. The primary difference between SBC Types I/II and SBC Type IV is that Type IV may have load-bearing walls constructed of wood.

_____

_____

_____ 7. Occupancy classifications are based on the premise that certain types of occupancies will naturally have higher fire loads and more occupants than other types of occupancies.

_____

_____

_____ 8. NFPA 101 includes as educational occupancies all buildings whose primary purpose is educational activities.

_____

_____

_____ 9. If a building has two different uses in distinctly different areas, both areas should be required to meet the requirements of the more restrictive occupancy rating.

_____

_____

_____ 10. If a building has different uses that are intermingled throughout the structure, the entire building should meet the requirements of the most stringent applicable occupancy rating.

_____

_____

Multiple Choice

C.   Write the letter of the best answer on the blank before each statement.

_____ 1. What NFPA standard addresses types of building construction?
   a. 101
   b. 220
   c. 120
   d. 201

_____ 2. In the NFPA construction classification Type IV-2HH, what does the *2* represent?

    a. The fire-resistive rating (in hours) of the exterior bearing walls

    b. The fire-resistive rating (in hours) of structural frames or columns and girders that support loads of more than one floor

    c. The fire-resistive rating (in hours) of the roof construction

    d. The fire-resistive rating (in hours) of the floor construction

_____ 3. In the NFPA construction classification Type I-443, what does the *3* represent?

    a. The fire-resistive rating (in hours) of the exterior bearing walls

    b. The fire-resistive rating (in hours) of structural frames or columns and girders that support loads of more than one floor

    c. The fire-resistive rating (in hours) of the roof construction

    d. The fire-resistive rating (in hours) of the floor construction

_____ 4. In the NFPA construction classification Type IV-2HH, what does the letter *H* indicate?

    a. Hundred

    b. Hours

    c. Heavy timber members

    d. Hardwood construction

_____ 5. What NFPA construction classification has interior structural members made of solid or laminated wood without concealed spaces?

    a. Type I   c. Type IV

    b. Type III   d. Type V

_____ 6. What NFPA construction classification has structural members including walls, columns, beams, floors, and roofs of noncombustible or limited-combustible materials?

    a. Type I   c. Type IV

    b. Type III   d. Type V

_____ 7. What NFPA construction classification has exterior walls and structural members of approved noncombustible or limited-combustible materials and interior structural members made wholly or partly of wood but smaller than required for heavy timber construction?

    a. Type II   c. Type IV

    b. Type III   d. Type V

____ 8. What NFPA construction classification has exterior walls, bearing walls, floors, roofs, and supports made completely or partially of wood with smaller dimensions than required for heavy timber construction?

    a. Type I   c. Type IV

    b. Type II   d. Type V

____ 9. What BOCA construction classification allows unprotected structural members?

    a. Type 1A   c. Type 2A

    b. Type 1B   d. Type 2C

____ 10. What additional BOCA construction classification allows unprotected structural members?

    a. Type 3A   c. Type 3B

    b. Type 2B   d. Type 1B

____ 11. Which of the following is not a subgroup for UBC ratings?

    a. P2   c. N

    b. F.R.   d. One hour

____ 12. Which UBC construction classification calls for structural elements of steel, iron, concrete, or masonry?

    a. Type I   c. Type III

    b. Type II   d. Type IV

____ 13. What two subgroups does the *Standard Building Code©* employ?

    a. F.R. and N

    b. Combustible and noncombustible

    c. One-hour protected and unprotected

    d. Extra protection and regular protection

____ 14. What size timbers does SBC Type III require for supporting roof and ceiling loads only?

    a. 6×6 inches *(150 × 150 mm)*   c. 8×8 inches *(200 × 200 mm)*

    b. 6×8 inches *(150 × 200 mm)*   d. 8×10 inches *(200 × 250 mm)*

____ 15. SBC Type V construction is also known as ____.

    a. Fire-resistive construction

    b. Noncombustible construction

    c. Ordinary construction

    d. Wood-frame construction

_____ 16. SBC Type VI construction is also known as _____.

    a. Fire-resistive construction

    b. Noncombustible construction

    c. Ordinary construction

    d. Wood-frame construction

_____ 17. What NFPA class of assembly occupancies can accommodate 300–1,000 persons?

    a. Class A   c. Class C

    b. Class B   d. Class D

_____ 18. How does NFPA 101 classify a university classroom?

    a. Place of assembly

    b. Educational occupancy

    c. Business occupancy

    d. Industrial occupancy

_____ 19. How does NFPA 101 classify a residential-custodial care facility?

    a. Place of assembly

    b. Health care occupancy

    c. Residential occupancy

    d. Business occupancy

_____ 20. What minimum number of residents does NFPA 101 require to classify a building as a detention and correctional facility?

    a. 4   c. 12

    b. 10   d. 25

_____ 21. How does NFPA 101 classify a college dormitory?

    a. Place of assembly

    b. Educational occupancy

    c. Residential occupancy

    d. Business occupancy

_____ 22. According to NFPA 101, which of the following is usually *not* a mercantile occupancy?

    a. Pharmacy

    b. Supermarket

    c. Shopping center

    d. Shopping mall

____ 23. How does NFPA 101 classify a college classroom building with rooms for fewer than 50 people?

    a. Place of assembly

    b. Business occupancy

    c. Educational occupancy

    d. Residential occupancy

____ 24. Which of the following does NFPA 101 *not* classify as an industrial occupancy?

    a. Instructional laboratory

    b. Dry-cleaning plant

    c. Refinery

    d. Telephone exchange

Identify

D. Identify the following abbreviations associated with construction and occupancy classifications. Write the correct interpretation before each.

_____ 1. NFPA

_____ 2. BOCA

_____ 3. ICBO

_____ 4. UBC

_____ 5. SBCCI

_____ 6. SBC

_____ 7. BCMC

Classify

E. Classify occupancies. Write the BOCA classification before each occupancy described below.

_____ 1. Performing arts center

_____ 2. Church

_____ 3. High school football stadium

_____ 4. Nursery school

_____ 5. Bakery

_____ 6. Brickyard

_____ 7. Building containing radioactive materials

_____ 8. Group home for eight mentally disabled residents

_____ 9. Clothing store

_____ 10. Dormitory

_____ 11. Detached one-family dwelling not more than three stories tall

_____ 12. Furniture warehouse

_____ 13. Meat locker

_____ 14. Carport

F. **Classify occupancies. Write the SBC classification before each occupancy described below.**

_____ 1. 950-seat television studio with a stage

_____ 2. Building for a parochial school

_____ 3. Soft drink bottling plant

_____ 4. Dynamite storage building

_____ 5. Tire warehouse

_____ 6. County jail with capacity for 25 inmates

_____ 7. Drugstore

_____ 8. Detached two-family dwelling

_____ 9. Book publisher's warehouse

_____   10. Wine cellar

_____   11. Water tower

G.   Classify occupancies. Write the UBC classification before each occupancy described below.

_____   1. Banquet hall with no stage and an occupant load greater than 300

_____   2. Outdoor rodeo arena

_____   3. Day care center for 12 children

_____   4. Machine shop

_____   5. Foundry

_____   6. Building where flammable liquids are stored under pressure

_____   7. Building where oxidizing cryogenic fluids are stored

_____   8. Computer chip manufacturing building

_____   9. Psychiatric hospital with 24 patients, 8 of whom are not capable of self-preservation

_____   10. Apartment building with four one-bedroom units

_____   11. Barn

_____   12. Enclosed parking garage

_____   13. Building where crates of bottled water are stored on pallets

_____   14. Water tower

Chapter 6 Site Access and Means of Egress

Matching

A. Match to their definitions terms associated with site access and means of egress. Definitions are continued on next page.

_____ 1. Continuous and unobstructed way of exit travel from any point in a building or structure to a public way

_____ 2. A street, alley, or parcel of land essentially open to the outside and used by the public

_____ 3. That portion of a means of egress that leads to the exit

_____ 4. That portion of a means of egress that is separated from the area of the building from which escape is to be made by walls, floors, doors, or other means that provide a protected path to the exterior of the building

_____ 5. That portion of a means of egress between the end of an exit and a public way

_____ 6. Stairway designed to limit the penetration of smoke, heat, and toxic gases

_____ 7. Means of egress to an area of refuge at approximately the same level in either the same building or another building

_____ 8. Ramp with a maximum slope of 1:10

_____ 9. Ramp with a maximum slope of 1:8

_____ 10. Total number of persons that may occupy a building or portion thereof at any one time

a. Occupant load

b. Exit

c. Exit discharge

d. Phase I evacuation

e. Public way

f. Smokeproof enclosure

g. Phase II evacuation

h. Egress

i. Horizontal exit

j. Exit access

k. *Life Safety Code®* Class B ramp

l. Phase III evacuation

m. *Life Safety Code®* Class A ramp

_____ 11. Evacuation of an entire floor of a health-
care facility, as wells as zones on the floor
above the fire floor

_____ 12. Evacuation of an entire zone of a health-
care facility

True/False

B.   **Write _TRUE_ or _FALSE_ before each of the following statements. Correct those statements that are false.**

_____ 1. Driveways and entrances should facilitate easy access for the largest fire apparatus expected to respond to an occupancy.

_____

_____

_____ 2. In some cases a fire inspector might be powerless to prevent the use of items that hinder firefighters' access to a structure's interior.

_____

_____

_____ 3. The Standard Building Code© specifies that a public way should have a width and height of at least 12 feet (4 m).

_____

_____

_____ 4. In general, doors should open in the direction of travel away from the exit.

_____

_____

_____ 5. Panic hardware is required for all exit doors.

_____

_____

_____ 6. The NFPA Life Safety Code® is the only model code that divides ramps into two classes.

_____

_____

_____ 7. Exit access corridors have stricter requirements than exit passageways.

_____

_____

_____ 8. In limited cases, the model building codes allow escalators to serve as part of a means of egress in existing buildings.

_____

_____

_____ 9. Fire escape ladders and slides are allowed only when approved by the AHJ.

_____

_____

_____ 10. Exit drills at educational facilities should emphasize discipline rather than speed.

_____

_____

_____ 11. Fire evacuation plans at correctional facilities must provide a nonelectrical means for releasing prisoners from their cells.

_____

_____

_____ 12. Inspection personnel may use hotel or motel employees as "fire wardens" to direct emergency evacuations.

_____

_____

_____ 13. Because occupants in hotels and motels are temporary, fire drills at these structures are futile.

_____

_____

Multiple Choice

C. **Write the letter of the best answer on the blank before each statement.**

_____ 1. The *Life Safety Code®* is _____.
    a. NFPA 101
    b. BOCA 200
    c. NFPA 1000
    d. ICBO 1A

_____ 2. When is access close to a building particularly important?
    a. When the nearest hydrant is more than 100 yards *(90 m)* from the building
    b. When the building contains no hazardous materials
    c. When aerial apparatus may be needed to ladder the building
    d. When the building has a high occupancy rating

_____ 3. Why is it important to note when a driveway or parking lot will not support the weight of a fully loaded apparatus?
    a. The apparatus will not be able to access the building.
    b. Damage to the pavement may lead to a lawsuit against the department.
    c. The owner can be ordered to replace the pavement.
    d. Drivers can take appropriate steps when positioning apparatus that require stabilizers.

_____ 4. How can personnel get a general idea of a driveway's construction if the owner does not know and construction plans are not available?
    a. Inspect the entire surface.
    b. Take a core sample.
    c. Consult codes current when the driveway was constructed.
    d. Observe the driveway's edge.

_____ 5. How should a jurisdiction deal with overhead obstructions?

    a. Inform driver/operators of their locations.

    b. Require corrections.

    c. Make corrections.

    d. Plan for their removal during an incident.

_____ 6. Who ordinarily has the authority to enforce fire lane parking laws?

    a. Property owners and police

    b. Property owners and fire marshals

    c. Fire marshals and fire inspectors

    d. Police and fire marshals

_____ 7. What should fire inspectors do when they find permanent changes in access to a structure's interior?

    a. Notify appropriate police department personnel.

    b. Notify appropriate fire department personnel.

    c. Immediately take steps to correct the situation.

    d. Notify appropriate building department personnel.

_____ 8. What should inspectors do when they find required exits, exit ways, or stairwells blocked?

    a. Notify appropriate police department personnel.

    b. Notify appropriate fire department personnel.

    c. Immediately take steps to correct the situation.

    d. Notify appropriate building department personnel.

_____ 9. Which of the following statements about code requirements for exit doors in new buildings is correct?

    a. All doors must be no more than 36 inches *(910 mm)* wide

    b. All doors must be at least 36 inches *(910 mm)* clear

    c. All doors must be no more than 42 inches *(1 060 mm)* wide

    d. All doors must be at least 32 inches *(810 mm)* clear

_____ 10. Which of the following statements about exit stairs is correct?

    a. Exit stairways must be at least 48 inches *(1 220 mm)* wide if the occupant load is 50 people or more.

    b. Landings must be provided to break up excessively long individual flights.

    c. Rails are required on only one side of stairs.

    d. All exit stairways must be at least 44 inches *(1 116 mm)* wide.

_____ 11. Which of the following statements about exit stairs is correct?

    a. If an exit stairway connects four or more stories, the separating construction must have at least a 2-hour fire-resistance rating.

    b. The separating construction for all exit stairways must have at least a 2-hour fire-resistance rating.

    c. Fire doors must have a 2-hour fire-resistance rating when used in a 2-hour rated enclosure.

    d. Exit stairs on new construction must be inside the building.

_____ 12. How is access made to a smokeproof enclosure?

    a. Directly through a fire door with a 2-hour fire-resistance rating

    b. Through a limited-access fire tunnel

    c. Through a vestibule or outside balcony

    d. Directly through a fire door with a 1-hour fire-resistance rating

_____ 13. Which of the following statements about horizontal exits is true?

    a. Horizontal exits may make up as much as 75 percent of a building's total exit capacity.

    b. The minimum fire-resistance rating for separating walls or partitions in horizontal exits is 1 hour.

    c. If a space used as an area of refuge includes a horizontal exit, it must have a standard exit that leads outside.

    d. A horizontal exit's area of refuge must only be large enough to shelter the occupants from the fire area.

_____ 14. Which of the following statements about BOCA, UBC, and SBC code requirements for ramps is true?

    a. Ramps serving occupant loads of 50 or more must be at least 48 inches *(1 220 mm)* wide.

    b. Slope requirements vary from code to code.

    c. Ramps serving occupant loads of less than 50 must be at least 44 inches *(1 116 mm)* wide.

    d. Exterior ramps generally do not have to provide as high a degree of protection as exterior stairs.

_____ 15. Probably the most important use of exit passageways is to _____.

    a. Provide access to exit access corridors

    b. Allow exit stairs to discharge directly outside multistory buildings

    c. Connect horizontal exits

    d. Provide an alternative for exit discharge

_____ 16. Which of the following statements about fire escape stairs is true?

    a. Fire escape stairs must be exposed to the fewest possible door or window openings.

    b. Fire escape stairs are permitted on new construction in limited circumstances.

    c. Fire escape stairs may constitute only one-fourth of the required means of egress in existing buildings.

    d. Windows may not be used as access to fire escape stairs in existing buildings.

_____ 17. Which of the following statements about exit illumination and marking is true?

    a. Ordinarily floors must be illuminated at not less than $1/5$ foot candle *(2.2 lx)*.

    b. Battery-operated units must be used for primary exit illumination.

    c. Floor-level exit signs may substitute for ceiling-level signs.

    d. Exit signs must be positioned so that no point in the exit area is more than 100 feet *(30.5 m)* from the nearest visible sign.

_____ 18. In most cases, how many exits do codes require from each story of a structure with an occupant load of 500 or less?

    a. 1   c. 3

    b. 2   d. 4

_____ 19. Which of the following best describes how exits should be located?

    a. As remote from each other as possible

    b. On opposite sides of hallways

    c. As close to each other as possible

    d. As far away from high hazard areas as possible

_____ 20. Which of the following statements regarding maximum travel distance to an exit is true?

    a. Travel distance refers to total time required to reach an exit.

    b. Maximum allowable travel distance generally remains the same whether a building is sprinklered or unsprinklered.

    c. The model codes all agree on maximum travel distance requirements.

    d. Travel distance is measured along the centerline of the natural path of travel.

_____ 21. What is typically one of an appointed evacuation monitor's duties?

    a. Extinguish fire with portable extinguisher.

    b. Rescue trapped coworkers.

    c. Inform coworkers of evacuation plan.

    d. Assist firefighters in entering the structure.

_____ 22. Who is responsible for making sure that an area is completely evacuated?

    a. Only the monitor

    b. All employees from that area

    c. All employees

    d. Only the employer

_____ 23. Who is responsible for planning and executing fire drills?

    a. Occupancy's fire loss prevention and control management staff

    b. Occupancy's designated emergency evacuation monitors

    c. Jurisdiction's fire and life safety education staff

    d. Jurisdiction's inspection staff

_____ 24. How often should most occupancies conduct fire drills?

    a. Once a year

    b. Twice a year

    c. Three times a year

    d. Four times a year

_____ 25. What should school officials do if weather conditions are likely to endanger students' health during the winter months?

    a. Schedule fewer fire drills.

    b. Proceed with regular fire drills without regard to weather.

    c. Reschedule missed fire drills for spring.

    d. Schedule weekly fire drills at the beginning of the school term.

_____ 26. What action is appropriate if a student cannot hold his or her place in a line moving at a reasonable speed?

    a. The entire line should slow down.

    b. A teacher or administrator should carry the student.

    c. The student should move independently of the regular line of march.

    d. Emergency evacuation wheelchairs should be kept on hand

_____ 27. Who should be responsible for searching rest rooms during emergency evacuations of schools?

    a. Student monitors

    b. All students

    c. Administrative personnel only

    d. Teachers and other staff members

___ 28. What acronym do many health-care facilities use for Phase I evacuation operations?

    a. ENFORCE   c. SAFETY

    b. PREVENT   d. REACT

___ 29. In a Phase IV evacuation, which area of a health-care facility is evacuated immediately following the fire floor?

    a. Floors above the fire floor

    b. First floor

    c. Floors below the fire floor

    d. All underground floors

___ 30. What causes most fires in correctional facilities?

    a. Careless use of smoking materials

    b. Cooking

    c. Arson

    d. Faulty electrical wiring

___ 31. What should fire personnel do first when asked to stand by at a major public event?

    a. Respectfully deny the request and suggest police personnel for the job.

    b. Check records of the occupancy's last inspection.

    c. Notify the fire marshal.

    d. Notify the fire chief.

Identify

D.   Identify the following abbreviations associated with fire codes. Write the correct interpretation before each.

_____ 1. AHJ

_____ 2. ADA

E.   Identify the responsibilities of fire inspectors or standby personnel at a major public event. Write an _X_ before each correct responsibility below.

_____ 1. Check any temporary electrical wiring used for the event.

_____ 2. Supervise crowd control.

_____ 3.   Ensure compliance with local liquor laws.

_____ 4.   Ensure that all exits are unlocked and exit lights are on.

_____ 5.   Supervise construction of emergency means of egress.

_____ 6.   Announce exit locations just before the event begins.

_____ 7.   Keep aisles open.

_____ 8.   Complete required private insurance forms.

_____ 9.   Enforce applicable smoking rules.

Chapter 7 Water-Based Fire Protection and Water Supply Systems

A. Match to their definitions terms associated with automatic sprinkler systems.

_____ 1. Sprinkler that protrudes downward from piping

_____ 2. Sprinkler that extends horizontally from piping along a wall

_____ 3. Device that prevents sprinkler water from backflowing into potable water supply

_____ 4. Section of pipe on which sprinklers are installed

_____ 5. Sprinkler system that contains water in the system at all times

_____ 6. Sprinkler system that maintains air under pressure in sprinkler piping

_____ 7. Device used to shut off water supply to entire system or zone

_____ 8. Device used to eliminate false alarms caused by pressure surges within the water supply system

_____ 9. Alarm-initiating device consisting of a vane that protrudes through the riser into the waterway

_____ 10. Device that simulates the operation of a single sprinkler and ensures that the alarm will operate even if only one sprinkler is fused in a fire

_____ 11. Type of sprinkler system used when preventing water damage is especially important

a. Wet-pipe

b. Preaction sprinkler system

c. Upright sprinkler

d. Sidewall sprinkler

e. Dry-pipe

f. Inspector's test connection

g. Check valve

h. Retard chamber

i. Branch line

j. Control valve

k. Water flow indicator

l. Pendant sprinkler

B. **Match to their definitions terms associated with fire pumps.**

_____ 1. The most common type of fire pump in fixed fire suppression systems

_____ 2. A pump whose main advantage is its compactness

_____ 3. Amount of pressure created by a column of water

_____ 4. A pump that is especially useful when the water source is below the pump

_____ 5. Source of power that operates a fire pump

_____ 6. Device that is designed to automatically start a stationary fire pump

_____ 7. Device designed to eliminate rapid opening and closing of an electric motor's main contacts

_____ 8. Graph showing a pump's performance at the factory

_____ 9. Gauge that measures velocity pressure of a flowing stream

_____ 10. Mathematical relationships used to correlate pump performance at any speed to its rated speed

a. Feet _(meters)_ of head

b. Running timer

c. Vertical shaft pump

d. Affinity laws

e. Certified shop test curve

f. Vertical turbine pump

g. Relativity principles

h. Pitot tube

i. Driver

j. Horizontal shaft pump

k. Controller

C. **Match to their definitions terms associated with water supply systems.**

_____ 1. Delivers water from source to distribution system without pumping

_____ 2. Network of water mains that makes up a distribution system

_____ 3. Fire hydrant that receives water from one direction

_____ 4. Fire hydrant that receives water from two or more directions

a. Fire flow test

b. Primary feeders

c. Grid

d. Distributors

e. Flow hydrant

f. Dead-end hydrant

_____ 5. Largest pipes in a distribution system

_____ 6. Pipes that reinforce the grid systems with a concentrated supply of water

_____ 7. Pipes that serve individual hydrants and blocks of customers

_____ 8. Hydrant with a base valve located below the frost line

_____ 9. Hydrant that has a compression valve at each outlet

_____ 10. Test to determine the rate of water flow available at various locations within a distribution system

_____ 11. Hydrant where static pressure and residual pressure are taken during a flow test

g. Circulating hydrant

h. Secondary feeders

i. Dry-barrel hydrant

j. Test hydrant

k. Wet-barrel hydrant

l. Gravity system

True/False

D. **Write _TRUE_ or _FALSE_ before each of the following statements. Correct those statements that are false.**

_____ 1. Sprinkler selection for a specific application should be based on the maximum temperature expected at the sprinkler's level.

_____

_____

_____ 2. The higher the RTI value, the faster a sprinkler device will operate.

_____

_____

_____ 3. Pressure tanks may be used as a primary water supply for residential sprinkler systems.

_____

_____

_____ 4. Fire department personnel should perform routine sprinkler system maintenance at industrial plants.

_____

_____

_____ 5. The only sure way to verify that a water tank is full is to climb to the top of the tank and peer into it.

_____

_____

_____ 6. For residential systems with two or more sprinklers, NFPA 13D requires an 18 gpm _(68 L/min)_ water supply for each sprinkler.

_____

_____

_____ 7. A ten-minute water supply in a residential sprinkler system prevents room flashover and provides for occupant safety.

_____

_____

_____ 8. Water mist systems may be suitable in many situations where halon was used previously.

_____

_____

_____ 9. Standpipe systems can take the place of automatic sprinklers in high-rise structures.

_____

_____

_____ 10. Class II standpipe systems are designed for use by building occupants who have no specialized fire training.

_____

_____

_____ 11. Class III standpipe systems have hose connections for use by both fire department personnel and the building occupants.

_____

_____

_____ 12. A wet standpipe system with limited water supply that keeps the system riser full is known as a "primed system."

_____

_____

_____ 13. Fixed fire pumps are rated at a fixed pressure.

_____

_____

_____ 14. When an electric horizontal split-case pump is tested, the measured voltage should never be more than 10 percent below the rated voltage.

_____

_____

_____ 15. Hydrant outlets are considered standard if there is at least one large outlet (4 or 4^1/$_2$ inches *[100 mm or 115 mm]*) and two outlets for 2^1/$_2$-inch *(65 mm)* couplings.

_____

_____

_____ 16. Encrustations are found chiefly in the bottom of water mains.

_____

_____

_____ 17. If flow pressure is inadequate to register on the pitot gauge, straight stream nozzles with orifices smaller than 2^1/$_2$ inches *(65 mm)* should be place on the hydrant outlet to increase flow velocity.

_____

_____

_____ 18. In general, when flow testing a single hydrant, the flow hydrant should be upstream from the test hydrant.

_____

_____

Multiple Choice

E. **Write the letter of the best answer on the blank before each statement.**

_____ 1. What percentage of fires do automatic sprinkler systems control or extinguish in buildings equipped with these systems?
   a. 94
   b. 96
   c. 98
   d. 100

2. What should a sprinkler or standpipe system include if the municipal water supply is not adequate to meet the system's demands?

   a. Auxiliary portable extinguishers and an on-site water supply

   b. An on-site water supply and additional fire exits

   c. On-site water supply and/or fire pumps

   d. Fire pumps and auxiliary portable extinguishers

3. Most fires are controlled with ____ or fewer sprinklers opening.

   a. 5   c. 15

   b. 10   d. 20

4. What color glass bulb sprinkler would best fit an application requiring a temperature rating of 325–375°F *(163°C to 191°C)*?

   a. Red   c. Blue

   b. Yellow   d. Purple

5. How far above the highest sprinkler in a the building should the bottom of a gravity tank be?

   a. 25 feet *(7.5 m)*   c. 35 feet *(11 m)*

   b. 30 feet *(9 m)*   d. 40 feet *(12 m)*

6. Clearance, the distance between the sprinkler and objects that could obstruct water distribution, should be measured from which part of the sprinkler head?

   a. Deflector   c. Valve cap

   b. Release mechanism   d. Lowest point on frame

7. How much clearance should be maintained under sprinklers?

   a. 6 inches *(150 mm)*   c. 18 inches *(450 mm)*

   b. 12 inches *(300 mm)*   d. 24 inches *(600 mm)*

8. If a sprinkler system includes 460 sprinklers, how many extra sprinklers, according to NFPA 13, need to be kept on hand?

   a. 30   c. 18

   b. 24   d. 12

9. How should fire inspectors check for obstructions in a sprinkler system's water supply?

   a. Visual inspection

   b. Flow test

   c. Pressure gauge reading

   d. Smoke test

_____ 10. How often must a wet-pipe system equipped with a fire pump be flow tested?

a. Daily

b. Weekly

c. Quarterly

d. Annually

_____ 11. How often should the heating system for a gravity tank and its piping be inspected during cold weather?

a. Daily

b. Weekly

c. Quarterly

d. Annually

_____ 12. How often should a wet-pipe system's main drain be inspected?

a. Daily

b. Weekly

c. Quarterly

d. Annually

_____ 13. When should low-point drains in a dry-pipe sprinkler system be tested?

a. Twice per year

b. Each fall

c. Each spring

d. Every three years

_____ 14. How often is a full-flow trip test recommended for a dry-pipe sprinkler system?

a. Twice per year

b. Each fall

c. Each spring

d. Every three years

_____ 15. How often should the supervisory air pressure in a preaction sprinkler system be checked?

a. Daily

b. Weekly

c. Monthly

d. Annually

_____ 16. How often should a preaction sprinkler system be trip tested?

a. Daily

b. Weekly

c. Monthly

d. Annually

_____ 17. Which of the following is the NFPA standard concerning sprinkler systems in one- and two-family homes?

a. NFPA 11D

b. NFPA 13D

c. NFPA 13R

d. NFPA 15R

_____ 18. What minimum water supply does NFPA 13R require for residential sprinkler systems?

a. 10-minute

b. 20-minute

c. 30-minute

d. 60-minute

_____ 19. Which of the following statements about residential sprinkler systems is correct?

    a. Residential sprinklers operate more quickly than conventional sprinklers.

    b. Residential sprinklers discharge water lower on the enclosing walls of a room than do standard sprinklers.

    c. Alarms for residential sprinklers are more complicated than those for standard systems.

    d. Residential sprinkler systems use the same valve arrangements as standard systems.

_____ 20. Which of the following statements about water mist systems is true?

    a. Water mist systems do not satisfactorily protect Class A combustibles.

    b. Water mist systems operate at lower pressures than standard sprinkler systems.

    c. Water mist systems are not suitable for protecting computer rooms.

    d. Water mist halts combustion by raising the humidity of the room to 100 percent.

_____ 21. What standard lists requirements for inspecting water mist systems?

    a. NFPA 1001   c. ISO 9000

    b. OSHA 129.b   d. NFPA 750

_____ 22. Which of the following statements about standpipe systems is true?

    a. Class II systems must be provided with 500 gpm *(1 893 L/min)* for at least 30 minutes.

    b. The required water supply for a standpipe system in a sprinklered structure is the sum of the sprinkler water demand and the standpipe system demand.

    c. Gravity tanks, manual fire pumps, and pressure tanks may supply water for standpipe systems.

    d. The Uniform Building Code requires a minimum residual pressure of only 65 psi *(448 kPa)*.

_____ 23. Which of the following is **not** an advantage of a primed wet standpipe system?

    a. Reduced minimum operating pressure

    b. Reduced time for water to reach hose valves

    c. Reduced water hammer

    d. Reduced corrosion of internal pipe surfaces

_____ 24. Which of the following is **not** a benefit of a pressure-regulating device in a standpipe system?

    a. Can prevent dangerous nozzle pressures

    b. Will simplify system design

    c. Will enhance system reliability

    d. May improve system economy

_____ 25. Which of the following statements about standpipe inspections is correct?

    a. Standpipe systems in new buildings should not be inspected until construction is complete.

    b. New systems should be hydrostatically tested at a minimum pressure of 200 psi *(1 379 kPa)* for one hour.

    c. All devices should be listed by a nationally recognized testing laboratory.

    d. On systems with automatic fire pumps, a flow test should be performed at the two lowest outlets.

_____ 26 How often should standpipe systems be visually inspected?

    a. Daily   c. Monthly

    b. Weekly   d. Semiannually

_____ 27. How often should dry standpipe systems be hydrostatically tested?

    a. Every year   c. Every three years

    b. Every two years   d. Every five years

_____ 28. Which of the following types of fire pump drivers is currently recognized by NFPA codes?

    a. Diesel engine   c. Natural gas engine

    b. Gasoline engine   d. Liquefied petroleum engine

_____ 29. What is an advantage of diesel engine drivers over electric motor drivers?

    a. Less expensive to operate

    b. Lower maintenance requirements

    c. Simpler operating principle and design

    d. Independent of local power supply

_____ 30. What device operates a fire pump's electric motor controller?

    a. Computer chip   c. Pressure-sensitive switch

    b. Thermostat   d. Alarm-activated switch

_____ 31. Which of the following statements regarding pump test manifolds is true?

    a. The test manifold piping must not have any valves.

    b. The hose valve header should be inside the pump room.

    c. Test manifold piping should be connected to the pump discharge line between the check valve and the indicating control valve.

    d. Pump installations equipped with a flowmeter rather than test headers and hose valves are not acceptable.

_____ 32. When testing pump installations, what percentage of rated net pressure must a pump develop while delivering 150 percent of the rated flow?

    a. 50   c. 75

    b. 65   d. 90

_____ 33. How much leakage is desirable in electric horizontal split-case pump shafts sealed with fiber packing?

    a. None   c. One drop per minute

    b. One drop per second   d. Two drops per second

_____ 34. How many gallons _(liters)_ per minute does a Class AA hydrant deliver?

    a. 1,500 _(5 678)_ or greater

    b. 1,000–1,499 _(3 785 to 5 677)_

    c. 500–999 _(1 893 to 3 784)_

    d. less than 500 _(1 893)_

_____ 35. What is the NFPA standardized color code for a Class B hydrant?

    a. Green   c. Orange

    b. Red   d. Light Blue

_____ 36. At minimum, how often should water main valves be operated?

    a. Every three months   c. Every year

    b. Every six months   d. Every two years

_____ 37. What type of water main valves are usually used in public water systems?

    a. PIV   c. OS&Y

    b. PIVA   d. Nonindicating

_____ 38. Who usually cleans water mains?

    a. Municipal water department

    b. Private company

    c. Fire department

    d. Municipal sewage department

_____ 39. Who is generally responsible for the repair and maintenance of fire hydrants on a public water system?

    a. Municipal water department

    b. Private company

    c. Fire department

    d. Municipal sewage department

_____ 40. At minimum, how often should fire flow tests be run on unchanged water supply systems?

    a. Every year   c. Every five years

    b. Every three years   d. Every ten years

_____ 41. In the formula $GPM = (29.83) \times C_d \times d^2 \times \sqrt{P}$ or $L/min = (0.0667766) \times C_d \times d^2 \times \sqrt{P}$, what is P?

    a. External pipe diameter   c. Pressure as read at the orifice

    b. Internal pipe diameter   d. Static pressure

_____ 42. To test available water supply, enough hydrants should be opened to reduce static pressure by at least _____.

    a. 5 percent   c. 20 percent

    b. 10 percent   d. 25 percent

Identify

F. Identify the following abbreviations associated with water-based protection and water supply systems. Write the correct interpretation before each.

_____ 1. RTI

_____ 2. OS&Y

_____ 3. PIV

_____ 4. WPIV

_____ 5. PIVA

G. Identify dry-pipe sprinkler system inspection concerns. Write an *X* before each correct statement below.

_____ 1. The drip ball valve should not allow water to seep from the fire department connection.

_____ 2. Any drum drips should be drained.

_____ 3. During freezing weather the dry-pipe valve enclosure heating device should keep the temperature of the dry-pipe valve at or above 40°F (4°C).

_____ 4. The system's air pressure should be maintained at 15–20 psi *(105 kPa to 140 kPa)* above the trip point.

_____ 5. Underground connections should be flushed following the same procedures used for wet-pipe sprinkler systems.

_____ 6. An air pressure loss of 0.75 psi *(5 kPa)* or greater during hydrostatic testing indicates leaks that must be located and corrected.

Chapter 8 Portable Fire Extinguishers, Special Agent Fire Extinguishing Systems, and Fire Detection and Alarm Systems

Matching

A. Match to their definitions terms associated with fire extinguishing agents.

_____ 1. Colorless, noncombustible gas that is heavier than air

_____ 2. Agent containing atoms from fluorine, chlorine, bromine, or iodine

_____ 3. Sodium bicarbonate, potassium bicarbonate, urea potassium bicarbonate, or potassium chloride

_____ 4. Monoammonium phosphate

_____ 5. Purple-K®

_____ 6. Dry powder for use on sodium and potassium fires but not suitable for magnesium fires

_____ 7. Dry powder for use on magnesium, sodium, and potassium fires

_____ 8. Dry powder consisting of a graphite base that conducts heat away from the fuel

a. Multipurpose dry chemical

b. Lith-X®

c. Potassium bicarbonate

d. Ordinary dry chemical agent

e. NA-X®

f. Carbon dioxide (CO_2)

g. Carbon monoxide (CO)

h. Halon

i. MET-L-X®

B. Match to their definitions terms associated with special agent fire extinguishing systems. Terms and definitions are continued on the next page.

_____ 1. Used where rapid fire knockdown is required and reignition is unlikely

_____ 2. Used to extinguish fires involving flammable liquids, gas, grease, or ordinary combustibles

a. Halogenated agent extinguishing system

b. High-expansion foam

_____ 3. Particularly useful in settings that require a "clean" extinguishing agent

_____ 4. CO_2 extinguishing system designed for very large hazards

_____ 5. Used for vapor suppression on unignited spills

_____ 6. Has an expansion ratio ranging from 20:1 to 200:1 and is especially useful in hard-to-reach spaces such as basements and other subterranean areas

_____ 7. Wetting agent that can penetrate bulk-bailed commodities such as paper, rags, and cardboard

_____ 8. Contains surfactants that allow it to shed or separate from hydrocarbon fuels

c. Wet-chemical extinguishing system

d. Foam solution

e. Dry-chemical extinguishing system

f. Low-pressure system

g. Low-expansion foam

h. Medium-expansion foam

i. Fluoroprotein foam

C. Match to their definitions terms associated with fire detection and alarm systems.

_____ 1. The "brain" of the alarm and detection system; often referred to as the alarm or annunciator panel

_____ 2. Usually consists of the building's main connection to the local public electric utility

_____ 3. Ensures that a system remains operational if main power supply fails

_____ 4. Intended solely to alert a building's occupants and ensure their life safety

_____ 5. Alerts emergency personnel before the general occupancy is notified

_____ 6. Auxiliary system that depends on the municipal fire alarm system's source of electrical current

a. Proprietary system

b. Presignal alarm

c. System control unit

d. Override system

e. Central station system

f. Remote station system

g. Local system

h. Secondary power supply

i. Parallel telephone system

_____ 7. Auxiliary system that connects each occupancy's alarm system to the fire department's dispatch center by an individual circuit

_____ 8. Transmits an alarm from the occupancy through connections or lines *other than* the municipal fire alarm box system

_____ 9. Each building or area of a complex has its own alarm system that is wired into a common receiving point somewhere in the facility

_____ 10. Each building or area of a complex has its own alarm system that is wired into a common receiving point at an outside, contracted service point

j. Shunt system

k. Primary power supply

D. **Match to their definitions terms associated with automatic alarm-initiating devices and testing fire detection and signaling systems. Terms and definitions are continued on the next page.**

_____ 1. Continuously monitors the atmosphere of an area for the products of combustion

_____ 2. The slowest type of system to activate under fire conditions

_____ 3. Under fire conditions, it melts and drops out, releasing a spring that completes an electrical connection

_____ 4. Under fire conditions, it breaks when the liquid it contains expands

_____ 5. Unlike "spot" detector types, it can be used to detect conditions over a wide area

_____ 6. Bonded metal strips that arc when heat causes one strip to expand faster than the other

_____ 7. Detects quick increases in temperature

a. Fusible link

b. Photoelectric smoke detector

c. Performance or service test

d. Continuous line detector

e. Fixed-temperature heat detector

f. Tube-type air sampling smoke detector

g. Frangible bulb

h. Rate-of-rise heat detector

_____ 8. Operates when smoke particles break a beam of light

_____ 9. Operates when combustion causes an electron imbalance among molecules

_____ 10. Test conducted when a fire detection and signaling system is installed

_____ 11. Test conducted periodically after a fire detection and signaling system is installed

i. Acceptance test

j. Bimetallic strip

k. Ionization smoke detector

l. Automatic alarm-initiating device

True/False

E. **Write** *TRUE* **or** *FALSE* **before each of the following statements. Correct those statements that are false.**

_____ 1. Class C Extinguishers are rated numerically from 1 to 40.

_____

_____

_____ 2. The manufacture of all new halon agents ceased on January 1, 1994.

_____

_____

_____ 3. Halon replacement agents such as FM 110 and Inergen (IG-541) are equally effective at the same volume as halon.

_____

_____

_____ 4. Dry chemicals can be projected more effectively from an extinguisher nozzle than can gaseous agents.

_____

_____

_____ 5. Pound for pound *(0.5 kg)*, potassium bicarbonate is about twice as effective as sodium bicarbonate as an extinguishing agent.

_____

_____

_____ 6. The primary advantage of pump-operated fire extinguishers is that their maintenance is extremely simple.

_____

_____

_____ 7. In ordinary occupancies, several lower-rated extinguishers may be used in place of one higher-rated extinguisher.

_____

_____

_____ 8. According to NFPA 10, any Class B fire in which the flammable liquid is $1/4$ inch *(6.4 mm)* or more deep is considered to be *with depth*.

_____

_____

_____ 9. Portable extinguishers installed outdoors can be protected with plastic bags.

_____

_____

_____ 10. A specialized extinguishing system is considered successful when it controls a fire.

_____

_____

_____ 11. Fire inspectors should be able to inspect special agent extinguishing systems for proper pressure on stored-pressure containers.

_____

_____

_____ 12. Storage containers for Halon 1301 extinguishing systems are tan colored.

_____

_____

_____ 13. Total flooding carbon dioxide extinguishing systems must be equipped with predischarge alarms.

_____

_____

_____ 14. Foams designed solely for hydrocarbon fires will not extinguish polar solvent fires, regardless of the concentration at which they are used.

_____

_____

_____ 15. A high-expansion foam system can fill an entire building to several feet *(meters)* above the highest storage area within a few minutes.

_____

_____

_____ 16. Discharging AFFF through standard sprinklers rather than foam sprinklers tends to undermine the spray and penetration.

_____

_____

_____ 17. Fire detection and alarm systems can be designed to shut down HVAC systems, close fire doors, pressurize stairwells, and automatically return elevators to a designated floor.

_____

_____

_____ 18. Remote stations must be monitored by fire department personnel.

_____

_____

_____ 19. A smoke detector senses fire much more quickly than does a heat detector.

_____

_____

_____ 20. Fire department personnel must participate in every performance or service test of a fire detection and signaling system.

_____

_____

Multiple Choice

F. **Write the letter of the best answer on the blank before each statement.**

_____ 1. Which of the following portable fire extinguishers is suitable for use on electrical fires?
   a. Class A   c. Class C
   b. Class B   d. Class D

_____ 2. Which of the following portable fire extinguishers is suitable for use on combustible metal fires?
   a. Class A   c. Class C
   b. Class B   d. Class D

_____ 3. Which of the following portable fire extinguishers is suitable for use on fires involving flammable or combustible liquids?
   a. Class A   c. Class C
   b. Class B   d. Class D

_____ 4. According to extinguisher tests, how much water does a water-type portable extinguisher rated 1-A require?
   a. 1 gallon *(3.79 L)*   c. $1^1/2$ gallons *(5.68 L)*
   b. $1^1/4$ gallons *(4.73 L)*   d. 2 gallons *(7.57 L)*

_____ 5. For what type fires would a portable extinguisher marked with a red square be suitable?
   a. Class A   c. Class C
   b. Class B   d. Class D

_____ 6. For what type fires would a portable extinguisher marked with a yellow star be suitable?

    a. Class A   c. Class C

    b. Class B   d. Class D

_____ 7. What size is the largest water extinguisher that can be considered portable?

    a. $1^1/4$ -gallon _(5 L)_   c. 5-gallon _(20 L)_

    b. $2^1/2$ -gallon _(10 L)_   d. 10-gallon _(40 L)_

_____ 8. How does carbon dioxide (CO_2) primarily extinguish a fire?

    a. Cooling   c. Chemical reaction

    b. Reducing fuel   d. Smothering

_____ 9. For what class(es) of fire would an AFFF portable extinguisher be suitable?

    a. Class A but not B or C   c. Classes A and B but not C

    b. Class B but not A or C   d. Classes A, B, and C

_____ 10. What is the primary advantage of halon extinguishers?

    a. Nontoxic   c. More effective than water on Class A fires

    b. Clean nature   d. Universally accepted

_____ 11. On what classes of fire is sodium bicarbonate **_most_** effective?

    a. Classes A and C   c. Classes A and B

    b. Classes B and C   d. Classes C and D

_____ 12. Which dry-chemical agent comes closest to being equally effective on all fires?

    a. Sodium bicarbonate   c. Potassium chloride

    b. Potassium bicarbonate   d. Monoammonium phosphate

_____ 13. Which dry-chemical agent melts and forms a solid coating on Class A materials?

    a. Sodium bicarbonate   c. Potassium chloride

    b. Potassium bicarbonate   d. Monoammonium phosphate

_____ 14. Which of the following statements about auxiliary fire-extinguishing equipment is true?

    a. Hanging buckets should have flat bottoms so that they will be easier to handle in an emergency.

    b. Buckets filled with sand are mainly used to extinguish incipient Class A fires.

    c. Flame-resistant wool is a suitable fiber for fire blankets.

    d. Fire blankets are of little use when a person's clothing catches fire.

____ 15. The general operating instructions for portable extinguishers follow the acronym
____.
a. PASS   c. PUMP
b. FIRE   d. SAFE

____ 16. Which of the following would be a light hazard occupancy?
a. Paint shop   c. School classroom
b. Mercantile display   d. Warehouse

____ 17. Which of the following would be an ordinary hazard occupancy?
a. Church   c. Restaurant with deep fat fryers
b. Automobile showroom   d. Office building

____ 18. Which of the following would be an extra hazard occupancy?
a. Auditorium   c. Parking garage
b. School laboratory   d. Boat maintenance facility

____ 19. Which of the following extinguisher ratings is required to protect a flammable
liquid fire hazard of 25 square feet *(2.32 m²)* with depth?
a. 10-B   c. 25-B
b. 20-B   d. 50-B

____ 20. Which of the following statements about portable extinguisher placement is true?
a. Extinguishers with a gross weight greater than 40 pounds *(18 kg)* should be
installed so that the top of the extinguisher is no more than 5 feet *(1.5 m)*
above the floor.
b. Ethylene glycol antifreeze may be used in plain water extinguishers.
c. The clearance between the bottom of the extinguisher and the floor should
never be less than 4 inches *(100 mm)*
d. Extinguishers should never be installed where temperatures are expected to
fall below freezing.

____ 21. Who is responsible for inspecting portable extinguishers?
a. Property owner or building occupant
b. Fire inspector
c. Extinguisher manufacturer
d. Building inspector

____ 22. What is the minimum weight by which an extinguisher may be deficient before it should be removed from service?

    a. 5 percent   c. 20 percent

    b. 10 percent   d. 25 percent

____ 23. Which of the following statements about extinguisher maintenance is correct?

    a. The fire inspector is responsible for reviewing all extinguisher maintenance records.

    b. Carbon dioxide hoses must be nonconductive.

    c. NFPA 10 requires all fire extinguishers to be thoroughly inspected at least twice a year.

    d. Stored-pressure extinguishers that require a twelve-year hydrostatic test must be emptied every three years for complete maintenance.

____ 24. How often should building representatives inspect pressure gauges and manual actuators on dry-chemical extinguishing systems?

    a. Weekly   c. Semiannually

    b. Monthly   d. Annually

____ 25. How often should dry-chemical extinguishing agents be inspected for obstructions, reduction of flow capabilities, and caking?

    a. Weekly   c. Semiannually

    b. Monthly   d. Annually

____ 26. At minimum, how often should fusible-link actuators in dry-chemical extinguishing systems be replaced?

    a. Weekly   c. Semiannually

    b. Monthly   d. Annually

____ 27. Who should inspect halon extinguishing systems?

    a. Any fire department inspector

    b. Any fire brigade member

    c. Only property maintenance staff

    d. Only specially trained personnel

____ 28. How often should halon extinguishing systems be inspected?

    a. Weekly   c. Semiannually

    b. Monthly   d. Annually

____ 29. How often should agent cylinders in CO_2 systems be inspected?

    a. Weekly   c. Semiannually

    b. Monthly   d. Annually

_____ 30. What proportion of foam concentrate do polar solvent fires require?

     a. 2 or 4 percent   c. 1 or 5 percent

     b. 3 or 6 percent   d. 2 percent

_____ 31. Which of the following is *not* a characteristic of aqueous film forming foam?

     a. Can be premixed

     b. Can be used in common fog nozzles

     c. Can be stored at temperatures below 32°F *(0°C)*

     d. Can penetrate high surface tension fuels

_____ 32. What foam proportioner has the ability to discharge foam from some outlets and plain water from others at the same time?

     a. Pressure proportioning tank system

     b. Around-the-pump

     c. Coupled water-motor pump

     d. Balanced pressure

_____ 33. What foam proportioner uses an in-line eductor on a pump-bypass line?

     a. Pressure proportioning tank system

     b. Around-the-pump

     c. Coupled water-motor pump

     d. Balanced pressure

_____ 34. In which of the following foam proportioners does water from the supply source displace water in the foam concentrate tank?

     a. Pressure proportioning tank system

     b. Around-the-pump

     c. Coupled water motor-pump

     d. Balanced pressure

_____ 35. Which of the following foam proportioners uses two positive-displacement rotary-gear pumps?

     a. Pressure proportioning tank system

     b. Around-the-pump

     c. Coupled water-motor pump

     d. Balanced pressure

_____ 36. Which of the following foam systems is a complete installation piped from a central foam station?

     a. Fixed   c. High-expansion

     b. Semifixed Type A   d. Foam/water

_____ 37. Which foam system uses a mobile foam solution apparatus?

    a. Fixed

    b. Semifixed Type A

    c. High-expansion

    d. Foam/water

_____ 38. How often should valves and alarms attached to foam extinguishing systems be inspected?

    a. Weekly

    b. Monthly

    c. Semiannually

    d. Annually

_____ 39. How often should a foam extinguishing system's foam concentrates, equipment, and proportioning systems be checked?

    a. Weekly

    b. Monthly

    c. Semiannually

    d. Annually

_____ 40. How soon after the main power supply's failure must the secondary system be able to make the detection and signaling system fully operational?

    a. 10 seconds

    b. 15 seconds

    c. 30 seconds

    d. 60 seconds

_____ 41. How long must a secondary power supply be able to supply a remote station system's maximum normal load?

    a. 5 minutes

    b. 60 hours

    c. 24 hours

    d. 2 hours

_____ 42. Which of the following statements about manual alarm-initiating devices is true?

    a. Fire alarm pull stations should be no less than $4^1/2$ feet *(1.3 m)* from the floor.

    b. Travel distances to the nearest pull station may not exceed 100 feet *(30 m)*.

    c. Multistory buildings should have at least two pull stations on each floor.

    d. Pull stations that require the operator to break a small piece of glass with a mallet are recommended.

_____ 43. Which of the following statements about flame detectors is true?

    a. Arc welding can trigger false alarms in ultraviolet detectors.

    b. They are relatively slow to respond to fire.

    c. Infrared detectors are ineffective in monitoring large areas.

    d. Infrared detectors are typically designed to respond to 1 square foot *(0.09 m²)* of fire from a distance of 25 feet *(7.5 m)*.

_____ 44. For fire protection purposes, it is only practical to monitor the levels of _____ and _____.

    a. Water vapor and carbon dioxide

    b. Hydrogen fluoride and hydrogen chloride

    c. Carbon monoxide and carbon dioxide

    d. Hydrogen cyanide and hydrogen sulfide

_____ 45. What task related to fire detection and signaling systems is generally a fire inspector's responsibility?

    a. Supervising system tests   c. Maintaining systems

    b. Operating systems   d. Performing system tests

_____ 46. Which of the following should *not* be used to check a restorable heat detector?

    a. Hair dryer   c. Heat lamp

    b. Open flame   d. Specialized testing appliance

_____ 47. What is the recommended way of cleaning dust or dirt from alarm initiating and indicating devices?

    a. Wiping   c. Feather dusting

    b. Blowing   d. Vacuuming

Identify

G. **Identify the following abbreviations associated with portable extinguishers, special agent extinguishing systems, and fire detection and alarm systems. Write the correct interpretation before each.**

_____ 1. AFFF

_____ 2. APW

_____ 3. PASS

_____ 4. FFFP

_____ 5. UV

_____ 6. IR

H. Identify fire detection and signaling equipment service testing procedures. Write an *X* before each correct statement below.

_____ 1. Wiring conduits should be inspected for solid connections and proper support.

_____ 2. Extra parts and testing equipment should be removed from control panels if storage space is not designed into the unit.

_____ 3. All alarm initiation devices should be activated during a performance test.

_____ 4. Detectors that have been painted should be cleaned and returned to service.

_____ 5. A permanent record of detector tests should be maintained for a minimum of 10 years.

_____ 6. Nonrestorable fixed-temperature detectors do not have to be tested until 15 years after their installation.

_____ 7. Pneumatic detectors should be tested semiannually.

_____ 8. Only fire inspectors should test flame and gas detection devices.

_____ 9. The same restorable heat detection device on each signal circuit should be tested semiannually.

_____ 10. NFPA codes require local alarm systems to be tested annually.

_____ 11. Manual fire alarm devices in central station systems should be tested semiannually.

_____ 12. The operation of auxiliary systems should be tested monthly.

_____ 13. The authority having jurisdiction establishes most testing requirements for remote station and proprietary systems.

_____ 14. All parts of emergency voice/alarm systems must be checked at least monthly.

Chapter 9 Plans Review

A. **Match to their definitions terms associated with plans review.**

_____ 1. Method of providing a view in exact proportion to the building's actual size

_____ 2. Two-dimensional drawing of a site or building as seen from directly above

_____ 3. The placing of a building on a site plan

_____ 4. Lines that indicate existing grade elevations

_____ 5. Lines that indicate planned elevations after grading is completed

_____ 6. Plan view that shows interior and exterior walls, doors and windows, ceiling joists, electrical outlets, and fixtures

_____ 7. Two-dimensional view of a building as seen from the exterior

_____ 8. Vertical view that shows an assembly's internal construction

_____ 9. The process of listing materials used to construct a wall

a. Site plan

b. Contour lines

c. Dimensioning

d. Grade lines

e. Calling up

f. Scaling

g. Elevation view

h. Floor plan

i. Plan view

j. Sectional view

9

B. **Write *TRUE* or *FALSE* before each of the following statements. Correct those statements that are false.**

_____ 1. Not all computer-aided drawings are drawn to scale.

_____

_____

_____ 2. A sectional view shows slopes that might affect placement of fire department ground ladders.

_____

_____

_____ 3. The left elevation is on the left side of the building when viewed from the rear.

_____

_____

_____ 4. The surrounding grades and the nature of intersections are beyond the scope of a fire inspector's review of drawings.

_____

_____

_____ 5. The site plan review may involve an assessment of the proposed water supply system.

_____

_____

_____ 6. The fire inspector is responsible for verifying that each means of egress provides a continuous path of travel to a safe refuge.

_____

_____

_____ 7. Automatic sprinkler systems do not affect requirements for interior finishes.

_____

_____

_____ 8. Exits must be illuminated at all times.

_____

_____

_____ 9. A designer's specifications for a special agent extinguishing system should include the concentration of extinguishing agent to be developed, the type of expellant gas, and the type of presignaling devices used if required.

_____

_____

_____ 10. General construction plans must include detailed information about fire alarm and communications systems.

_____

_____

_____ 11. Some jurisdictions allow manual pull stations to be omitted when occupancies are protected by sprinkler systems that sound an alarm when activated.

_____

_____

Multiple Choice

C. **Write the letter of the best answer on the blank before each statement.**

_____ 1. What are the basic tools of a plans reviewer?
  a. Fines and penalties
  b. Drafting skills
  c. Interpersonal skills
  d. Fire/building codes and standards

_____ 2. What part of a working drawing includes the scale of the drawing, name of the firm producing the drawing, and name of the person performing the drafting?

    a. Nameplate   c. Fact sheet

    b. Title block   d. Copyright block

_____ 3. What part of a working drawing shows _when_ it was drawn?

    a. Revisions block   c. Time stamp

    b. Dateline   d. Modifications block

_____ 4. What does the letter _E_ indicate in a drawing marked "E3"?

    a. Drawn by a licensed engineer

    b. Level of energy efficiency

    c. Drawn in sequence after drawings marked "D"

    d. Electrical sheet

_____ 5. Which of the following plans shows information needed to locate a building?

    a. Topography   c. Site

    b. Survey   d. Plot

_____ 6. Which of the following plans shows the location of utility lines?

    a. Site   c. Ceiling

    b. Floor   d. Roof

_____ 7. How are structures that must be removed from the area shown on a site plan?

    a. Dotted line   c. Solid line

    b. Broken line   d. Double lines

_____ 8. Which of the following drawings shows the location of fire hydrants and water mains?

    a. Site plan   c. Elevation view

    b. Floor plan   d. Detail view

_____ 9. What is the usual symbol for walls on a floor plan?

    a. Solid line   c. Dotted line

    b. Double broken line   d. Parallel solid lines

_____ 10. What indicates the swing of a door on a floor plan?

    a. Arrow   c. Arc

    b. Doorknob   d. Circled letter _L_ or _R_

_____ 11. Which of the following drawings usually shows window sizes?

      a. Floor plan   c. Sectional view

      b. Elevation view   d. Detail view

_____ 12. Which of the following drawings shows width and thickness of footings?

      a. Site plan   c. Elevation view

      b. Floor plan   d. Sectional view

_____ 13. Which of the following individuals would most likely conduct formal plans reviews for fire code enforcement?

      a. Fire inspector   c. Fire chief

      b. Building engineer   d. Fire protection engineer

_____ 14. What should a fire inspector do first when reviewing construction documents?

      a. Locate hydrants and water mains.

      b. Evaluate automatic fire protection systems.

      c. Review the building's overall size; i.e., height and area.

      d. Locate fire department connections.

_____ 15. Who should determine the occupancy classification for a proposed building?

      a. Architect   c. Fire marshal

      b. Fire inspector   d. Building inspector

_____ 16. After verifying that a project will be built, what should fire inspectors determine first?

      a. Life safety features

      b. Means of egress

      c. Fire protection features

      d. Occupant load

_____ 17. Which of the following will help an inspector determine if escalators are adequately protected?

      a. Floor plan and elevation view

      b. Electrical and mechanical plans

      c. Sectional and mechanical plans

      d. Electrical plan and elevation view

_____ 18. Which of the following statements regarding HVAC plan reviews is true?

    a. Ductwork may not penetrate fire-resistive walls or floors.

    b. Exits may be used for supply or return air.

    c. Smoke control systems are usually used only in buildings involving large numbers of people.

    d. Total exhaust mode can be crucial in removing heat and smoke from a building during fire conditions.

_____ 19. What determines which standard applies to a building's sprinkler system?

    a. Location and value

    b. Occupancy and storage commodity

    c. Occupancy and value

    d. Storage commodity and value

_____ 20. Which of the following statements regarding the review of automatic sprinkler system plans is true?

    a. A sprinkler system's size is based on the total area protected by one pair of risers and one control valve.

    b. The fire inspector should provide a graph that compares water demand to available water supply.

    c. Flow tests are used to measure the pressure and flow available at the control valve.

    d. Fire inspectors should check the total capacity of the water supply.

_____ 21. From the building occupant's point of view, which of the following alarm systems is most important?

    a. Local   c. Remote station

    b. Auxiliary   d. Proprietary

_____ 22. Which of the following buildings would most likely require an annunciator panel?

    a. Single-family residence   c. Hospital

    b. Convenience store   d. Two-car detached garage

_____ 23. Which of the following drawings should fire inspectors use to verify the location and number of alarm-actuating and signaling devices?

    a. Elevation views

    b. Floor plans

    c. Sectional views

    d. Detail views

D. Identify the following abbreviations associated with plans review. Write the correct interpretation before each.

_____ 1. CAD

_____ 2. HVAC

Chapter 10 Identification of Hazardous Materials and Storage, Handling, and Use of Flammable and Combustible Liquids

Matching

A. Match to their class types of hazardous materials.

_____ 1. Oxidizers   a. Class 1

_____ 2. Corrosives   b. Class 2

_____ 3. Flammable liquids   c. Class 3

_____ 4. Radioactive substances   d. Class 4

_____ 5. Explosives   e. Class 5

_____ 6. Poisons and infectious substances   f. Class 6

_____ 7. Flammable solids   g. Class 7

_____ 8. Gases   h. Class 8

                                                    i. Class 9

B. Match to their class and division types of hazardous materials. Terms and definitions are continued on next page.

_____ 1. Poisonous material   a. 1.2

_____ 2. Poisonous gas   b. 1.4

_____ 3. Infectious substance   c. 2.2

_____ 4. Explosives that present a minor explosion hazard   d. 2.3

                                                    e. 3.3

_____ 5. Spontaneously combustible materials

                                                    f. 4.1

_____ 6. Nonflammable compressed gas

_____ 7. Explosives that present a projection hazard but not a mass explosion hazard

g. 4.2

h. 5.1

_____ 8. Flammable solid

i. 6.1

_____ 9. Oxidizer

j. 6.2

C. **Match to their section numbers classes of information on an MSDS.**

_____ 1. Reactivity data

a. Section I

_____ 2. Hazardous ingredients

b. Section II

_____ 3. Control measures

c. Section III

_____ 4. Fire and explosion hazard data

d. Section IV

_____ 5. Health hazard data

e. Section V

_____ 6. Precautions for safe handling and use

f. Section VI

_____ 7. Physical and chemical characteristics

g. Section VII

h. Section VIII

D. **Match to their definitions terms related to flammable and combustible liquids.**

_____ 1. Liquids having a flash point below 100°F _(38°C)_ and a vapor pressure not exceeding 40 psi _(256 kPa)_

a. Flash point

b. Specific gravity

_____ 2. Liquids having a flash point at or above 100°F _(38°C)_

c. Flammable and explosive limits

_____ 3. Minimum temperature at which a liquid fuel gives off sufficient vapors to form an ignitable mixture with the air near the surface

d. Boiling point

e. Combustible liquids

_____ 4. Temperature at which a liquid fuel, once ignited, will continue to burn

f. Autoignition temperature

_____ 5. Temperature at which a fuel will ignite independent of another ignition source

g. Fire point

h. Flammable liquids

10

6. Temperature at which the vapor pressure of a liquid is equal to the external pressure applied to it

7. Ratio of a liquid's weight to the weight of an equal volume of water

True/False

E. Write *TRUE* or *FALSE* before each of the following statements. Correct those statements that are false.

1. The best source for information on a specific hazardous material is the MSDS.

2. An MSDS must follow a set form specified by federal regulations.

3. An MSDS from the United States is acceptable for use in Canada.

4. Flash points determined by the open-cup method are approximately 10°F to 15°F *(3°C to 4°C)* higher than those determined by the closed-cup method.

5. Explosions of flammable vapor-air mixtures usually occur in a confined space such as a building, room, or container.

109

Fire Inspection and Code Enforcement

_____ 6. Distances from storage tanks to property lines and public ways are determined by federal statute.

_____

_____

_____ 7. Fusible vents on portable tanks must be designed to operate at a temperature not exceeding 212°F *(100°C).*

_____

_____

_____ 8. Loading and unloading stations for Class I liquids should be no closer than 25 feet *(8 m)* from property lines or adjacent buildings.

_____

_____

Multiple Choice

F. Write the letter of the best answer on the blank before each statement.

_____ 1. What resource lists the four-digit identification numbers assigned to hazardous materials?
   a. *United Nations Codebook*
   b. *IGGRN*
   c. NATO Haz Mat Treaty
   d. *NAERG*

_____ 2. The NFPA 704 system is designed to be used for _____.
   a. Manufacturing
   b. Transportation
   c. Chronic exposures
   d. Nonemergency occupational exposures

_____ 3. What does the number *4* signify in the NFPA 704 system?
   a. No hazard   c. Moderate hazard
   b. Low hazard   d. Severe hazard

10

_____ 4. What color indicates health hazard ratings in the NFPA 704 system?
    a. Red   c. Yellow
    b. Blue   d. White

_____ 5. What position indicates flammability ratings in the NFPA 704 system?
    a. 12 o'clock   c. 6 o'clock
    b. 3 o'clock   d. 9 o'clock

_____ 6. What color indicates reactivity hazard rating in the NFPA 704 system?
    a. Red   c. Yellow
    b. Blue   d. White

_____ 7. What position indicates special hazards in the NFPA 704 system?
    a. 12 o'clock   c. 6 o'clock
    b. 3 o'clock   d. 9 o'clock

_____ 8. Which section of an MSDS includes hazardous decomposition or by-products?
    a. Physical and chemical characteristics
    b. Hazardous ingredients
    c. Reactivity data
    d. Health hazard data

_____ 9. Which section of an MSDS includes emergency and first aid procedures?
    a. Precautions for safe handling and use
    b. Hazardous ingredients
    c. Health hazard data
    d. Physical and chemical characteristics

_____ 10. Which section of an MSDS lists a product's evaporation rate?
    a. Fire and explosion hazard data
    b. Hazardous ingredients
    c. Reactivity data
    d. Physical and chemical characteristics

_____ 11. Which class of flammable or combustible liquids has a flash point below 73°F _(23°C)_ and a boiling point at or above 100°F _(38°C)?_
    a. IB
    b. IC
    c. II
    d. IIIA

_____ 12. Which class of flammable or combustible liquids has a flash point at or above 140°F *(60°C)* and below 200°F *(93°C)?*

    a. IB   c. II

    b. IC   d. IIIA

_____ 13. Which of the following has a storage capacity of 60 gallons *(240 L)* or less?

    a. Fixed vessel   c. Portable tank

    b. Container   d. Storage tank

_____ 14. Which of the following has a storage capacity of more than 60 gallons *(240 L)* and remains in a fixed location?

    a. Fixed vessel   c. Portable tank

    b. Container   d. Storage tank

_____ 15. Which of the following storage tank types is designed for pressures of 0.5 to 15 psi *(3 kPa to 103 kPa)?*

    a. High-pressure   c. Low-pressure

    b. Medium-pressure   d. Atmospheric

_____ 16. Which of the following materials is suitable for aboveground storage tanks?

    a. Steel   c. Aluminum

    b. PVC   d. Ceramic

_____ 17. How far apart should tanks be if they contain unstable flammable or combustible liquids?

    a. At least 3 feet *(1 m)*

    b. Approximately 10 feet *(3 m)*

    c. Approximately one-sixth the sum of their diameters

    d. At least one-half the sum of their diameters

_____ 18. How far away should LPG containers be from flammable liquid storage tanks?

    a. 3 feet *(1 m)*

    b. 6 feet *(2 m)*

    c. 10 feet *(3 m)*

    d. 20 feet *(6 m)*

_____ 19. From a fire protection standpoint, which of the following storage tanks is the safest form of storage for flammable and combustible liquids?

    a. Aboveground   c. Underground

    b. Elevated   d. Inside a building

____ 20. Which of the following materials is *not* suitable for underground storage tanks?

    a. Metal
                  c. Unlined concrete

    b. Fiberglass
            d. Vitreous clay

____ 21. If vehicles are likely to pass over underground tanks, the tanks must be protected by ____.

    a. 2 feet *(0.6 m)* of earth

    b. 1 foot *(0.3 m)* of earth plus 4 inches *(100 mm)* of reinforced concrete

    c. 1.5 feet *(0.5 m)* of well-tamped earth plus 6 inches *(150 mm)* of reinforced concrete

    d. 3 feet of earth *(1 m)* plus 8 inches *(200 mm)* of asphaltic concrete

____ 22. How high above the adjacent ground level must vents be for underground tanks containing Class I liquids?

    a. 3 feet *(1 m)*
             c. 9 feet *(3 m)*

    b. 6 feet *(2 m)*
            d. 12 feet *(4 m)*

____ 23. Where should fill pipes for Class IB and IC liquids terminate?

    a. Within 6 inches *(150 mm)* of the tank bottom

    b. At least 12 inches *(300 mm)* from the tank bottom

    c. Within 6 inches *(150 mm)* of the top of the tank

    d. At least 6 inches *(150 mm)* from the top of the tank

____ 24. What causes electrolytic corrosion of underground storage tanks?

    a. Static electricity

    b. Using two different metals for tanks and piping

    c. Corrosive soils

    d. Leaking Class III liquids

____ 25. What is the maximum amount of Class II liquids that may be stored in a dwelling occupancy containing three or fewer dwelling units?

    a. 60 gallons *(240 L)*
        c. 10 gallons *(40 L)*

    b. 25 gallons *(100 L)*
       d. 20 gallons *(80 L)*

____ 26. What is the basic principle of a VRS?

    a. Chemical reaction between gasoline vapor and inert gas

    b. Closed-loop exchange of gasoline and vapor

    c. Decompression of gasoline vapor

    d. Secondary chamber containment of vapors

G. **Identify the following abbreviations associated with hazardous materials and with flammable and combustible liquids. Write the correct interpretation before each.**

_____ 1. DOT

_____ 2. MSDS

_____ 3. ORM-D

_____ 4. NAERG

_____ 5. CAS

_____ 6. LFL

_____ 7. UFL

_____ 8. BLEVE

_____ 9. VRS

_____ 10. EPA

Chapter 11 Storage, Handling, and Use of Other Hazardous Materials

Matching

A. **Match to their definitions types of magazines (storage facilities for explosives).**

_____ 1. Permanent, portable, or mobile magazine constructed of masonry that will not mass detonate when initiated by a No. 8 test blasting cap

_____ 2. Portable facility for temporary storage of explosive materials under the constant attendance of a qualified employee

_____ 3. Portable or mobile magazine used for outdoor or indoor storage of explosives that will detonate when initiated by a No. 8 test blasting cap

_____ 4. Permanent facility for the storage of explosives that will detonate when initiated by a No. 8 test blasting cap

a. Type 1 magazine

b. Type 2 magazine

c. Type 3 magazine

d. Type 4 magazine

e. Type 5 magazine

B. **Match to their definitions terms related to hazardous materials. Terms and definitions are continued on the next page.**

_____ 1. Liquids, solids, or gases that even in small quantities and without an external ignition source can ignite within five minutes after coming into contact with air

_____ 2. Substances that have the potential to self-heat when in contact with air and without an energy supply

_____ 3. Substances capable of producing serious illness or death once they enter the bloodstream.

a. Chronic toxicity

b. Corrosives

c. Beta radiation

d. Self-heating materials

e. Dangerous-when-wet materials

_____ 4. The concentration of a given toxic material that generally may be tolerated without ill effects.

_____ 5. Effects of a substance upon repeated exposures over a long time

_____ 6. Radiation that a fairly thin layer of metal or plastic can stop

_____ 7. Radiation that only lead or concrete can stop

_____ 8. Measurement of radiation doses

f. Pyrophoric materials

g. Toxic or highly toxic materials

h. Gamma radiation

i. Roentgen

j. Threshold limit value

True/False

C. **Write** _TRUE_ **or** _FALSE_ **before each of the following statements. Correct those statements that are false.**

_____ 1. In normal storage and transport states, explosives are extremely unstable and unsafe to handle.

_____

_____

_____ 2. Gases with a vapor density greater than 1.0 will rise and concentrate near the ceiling.

_____

_____

_____ 3. If gas is transferred in a completely closed system, the system need not be electrically grounded or bonded.

_____

_____

_____ 4. Local jurisdictions may enact regulations in addition to federal requirements for transporting compressed and liquefied gases.

_____

_____

_____ 5. During storage and handling, oil should not be placed in containers with any type of flammable solid.

_____

_____

_____ 6. LD_{50} refers to the air concentration of a given substance that was lethal to 50 percent or more of test animals when they inhaled the substance.

_____

_____

_____ 7. Inspectors generally do not have the authority to ask to see verification that employees are adequately trained to work with toxic materials.

_____

_____

_____ 8. Oxidizers are not necessarily combustible by themselves.

_____

_____

_____ 9. Occupancies that handle radioactive materials should have an alarm panel in the area of the facility where the materials are stored.

_____

_____

Multiple Choice

D. Write the letter of the best answer on the blank before each statement.

_____ 1. Which agency issues regulations for the manufacture, distribution, and storage of explosives?
  a. OSHA
  b. DOT
  c. BATF
  d. EPA

_____ 2. Which of the following statements about storing explosives is true?
  a. The brand of an explosive is not relevant.
  b. Piled oxidizers must be separated from readily combustible fuels.
  c. Dynamite must be stored on end.
  d. Packages must be opened or repackaged at least 100 feet *(30 m)* from the magazine.

_____ 3. How often should magazines be opened and inspected?
  a. Daily
  b. Every three days
  c. Every five days
  d. Weekly

_____ 4. What minimum distance should be maintained between vehicles transporting explosives?
  a. 100 feet *(30 m)*
  b. 200 feet *(60 m)*
  c. 300 feet *(90 m)*
  d. 500 feet *(150 m)*

_____ 5. How long after they arrive at a terminal may explosives remain there?
  a. 2 hours
  b. 12 hours
  c. 24 hours
  d. 48 hours

_____ 6. Which of the following is an industrial gas?
  a. Argon
  b. Propane
  c. Phosphine
  d. Cyclopropane

_____ 7. Which of the following is a medical gas?
  a. Argon
  b. Propane
  c. Phosphine
  d. Cyclopropane

_____ 8. According to _____ , the volume of a confined gas varies inversely with the applied pressure.

    a. Boyle's Law   c. Charles' Law

    b. Lane's Law   d. Taylor's Law

_____ 9. What organization sets standards for tanks designed to hold large quantities of low-pressure gas?

    a. ASME   c. CTC

    b. API   d. ISO

_____ 10. Which of the following statements about storage cylinders for compressed and liquefied gases is true?

    a. Cylinders should be stored horizontally on a level floor.

    b. Gas cylinders are designed for temperatures no greater than 130°F *(54°C)*.

    c. Empty cylinders should be stored in the same area with full cylinders.

    d. Grass, weeds, and other combustibles should not be within 20 feet *(6 m)* of the tank.

_____ 11. What is the primary hazard associated with the storage of compressed and liquefied gases?

    a. High toxicity   c. High reactivity

    b. High flammability   d. High pressure

_____ 12. Which of the following is *not* a basic cause of container failure that leads to uncontrolled release of energy?

    a. Static electricity within the container

    b. Excessive pressure within the container

    c. Flame impingement on the container

    d. Mechanical damage to the container

_____ 13. When gas cylinders are nested without support, how many points of each cylinder must be in contact with other cylinders?

    a. One   c. Three

    b. Two   d. Four

_____ 14. What is the symbol for lift trucks rated for use in atmospheres containing flammable vapors?

    a. LX   c. ELP

    b. LF   d. EX

_____ 15. Which of the following statements about handling gas cylinders is true?

    a. Lift cylinders with adequately rated electromagnets.

    b. Do not roll cylinders on their bottom edges.

    c. Do not allow hydrostatic testing numbers to be changed.

    d. When cylinders are empty, mark them "MT."

_____ 16. How often should the loading hose at a gas-cylinder loading station be replaced?

    a. Every ten years   c. Every two years

    b. Every five years   d. Every year

_____ 17. Which of the following statements about gas cylinders is correct?

    a. Oil or grease on oxygen cylinders can cause an explosion.

    b. Cylinders should be stored below ground level.

    c. Nonreturnable containers should not be punctured before disposal.

    d. Loading hoses with vaportight connections must be bonded.

_____ 18. What document details regulations for transporting compressed and liquefied gases?

    a. CFR 49   c. ISO 9000

    b. DOT 1970   d. NFPA 94

_____ 19. Most codes require that toxic material be separated from other hazardous materials by a least a _____ fire separation.

    a. One-half-hour-rated   c. Two-hour-rated

    b. One-hour-rated   d. Four-hour-rated

_____ 20. Which of the following statements about oxidizers is correct?

    a. Liquid- or slurry-form oxidizers are shipped in aluminum tank trucks.

    b. Organic peroxides may be shipped in unlimited quantities.

    c. When exposed to heat, oxidizers decompose, releasing oxygen that accelerates a fire.

    d. Hydrogen peroxide solutions may be shipped in tank cars made of stainless steel.

_____ 21. How many millirems of radiation may workers receive per year in the United States?

    a. 400   c. 1/1000

    b. 5,000   d. 1,000

_____ 22. What is the principle hazard associated with corrosives?

    a. Leakage   c. Fire

    b. Explosion   d. Smoke damage

_____ 23. Which of the following corrosives is noncombustible but may cause certain materials to spontaneously ignite?

    a. Iodine   c. Strong bases

    b. Bromine   d. Fluorine

_____ 24. What material is stored in wax-coated bottles because it attacks glass?

    a. Sulfuric acid   c. Perchloric acid

    b. Nitric acid   d. Hydrofluoric acid

_____ 25. If possible, what is the best way for fire inspectors to deal with ORM-D materials?

    a. Examine each material on an individual basis.

    b. Apply regulations from as many other categories as possible.

    c. Fit the material into a category, and apply those regulations.

    d. Treat like flammable and combustible materials.

Identify

E. **Identify the following abbreviations associated with hazardous materials. Write the correct interpretation before each.**

_____ 1. BATF

_____ 2. DOT

_____ 3. MEP

_____ 4. MAPP

_____ 5. CTC

_____ 6. CFR

_____ 7. TLV

_____ 8. RSO

_____ 9. NRC

F.   **Identify characteristics of cryogenic liquids. Write an *X* before each correct statement below.**

_____   1.   The increased weight of cryogenic liquids offsets their savings in storage space.

_____   2.   Cryogenic liquids are valuable simply because they are extremely cold.

_____   3.   One advantage of cryogenic liquids is their tremendous liquid-to-vapor ratio.

_____   4.   All cryogenic liquids except oxygen are either asphyxiants or toxic.

_____   5.   Small amounts of cryogenic liquids move quickly across the skin.

_____   6.   The inner tank of cryogenic-liquid containers is made of carbon steel.

_____   7.   Self-refrigeration maintains a uniform vapor pressure inside cryogenic containers.

_____   8.   The most common types of pressure relief devices for cryogenic containers are pressure relief valves, frangible disks, and safety vents.

_____   9.   All pipes should be sloped down from cryogenic containers.

Chapter 1 Answers

Matching

A.

1. f *(6)*	7. j *(12)*
2. c *(7)*	8. b *(18)*
3. i *(8)*	9. k *(18)*
4. m *(8)*	10. d *(18)*
5. a *(8)*	11. h *(19)*
6. l *(9)*	12. g *(19)*

True/False

B.

1. True *(5)*
2. False. Personnel other than designated fire inspectors *may perform inspection procedures. (5)*
3. False. Inspectors must *not* assume that statutes in place in their jurisdiction are the same for another jurisdiction. *(6)*
4. False. Fire inspectors who have the authority to issue summonses, write tickets, issue warrants, or make arrests *must have appropriate law enforcement training. (6)*
5. True *(7)*
6. True *(9)*
7. False. Smaller jurisdictions *may contract with an engineering or fire protection firm* to provide technical assistance when required. *(10)*
8. True *(10)*
9. True *(11)*
10. False. In most jurisdictions *fire inspectors themselves have little authority to approve modification of code requirements.* The inspector usually processes the requests and receives a formal interpretation from a superior inspection bureau officer, a staff fire protection engineer, a contract fire protection consultant, or an appeals board. *(12)*
11. True *(13)*
12. False. Complaints that do *not* require immediate action *can be routed through the normal channels for processing and handling. (14)*
13. False. In many situations it is *best not to give advance notice* of a complaint inspection. *(14, 15)*
14. True *(15, 16)*
15. False. Having a *good exchange of information with other agencies is a positive step* for keeping current with codes in a community or jurisdiction. *(16)*
16. True *(16)*
17. False. In most cases the local fire inspector is *not responsible* for enforcing federal regulations, but he/she should know how to report hazards or violations to the proper authority. *(17)*
18. True *(17)*
19. True *(17, 18)*
20. True *(18)*
21. False. In general, permits are *not* issued to allow the party to disregard or exceed code requirements in any manner. *(19)*
22. True *(19)*
23. True *(20)*
24. True *(20)*

25. False. The permit must *explicitly explain the conditions under which it has been issued.* This includes what actions are being allowed, guidelines that must be followed, and the time frame for which the permit is applicable. *(20)*
26. True *(20)*
27. True *(20)*
28. False. Depending on the problem, a permit *may be revoked permanently or temporarily* until the condition is corrected. *(20)*

E. *(10, 11)*

1, 3, 5, 7, 8

F. *(13, 14)*

3, 5, 7, 8, 9, 10

Multiple Choice

C.

1.	b *(5)*	13.	b *(16, 17)*
2.	d *(6)*	14.	a *(17, 18)*
3.	c *(6)*	15.	d *(18)*
4.	b *(7)*	16.	b *(19)*
5.	a *(11, 12)*	17.	c *(20)*
6.	d *(12)*	18.	b *(20)*
7.	b *(12)*	19.	a *(20)*
8.	c *(13)*	20.	d *(22)*
9.	a *(13)*	21.	c *(22)*
10.	d *(15)*	22.	a *(22)*
11.	b *(15)*	23.	c *(22, 23)*
12.	c *(16, 17)*		

Identify

D.

1. National Fire Protection Association *(18)*
2. Building Officials & Code Administrators International *(18)*
3. International Conference of Building Officials *(18)*
4. Southern Building Code Congress International *(18)*
5. Material safety data sheet *(20)*

Chapter 2 Answers

Matching

A.

1. d *(35)*
2. a *(35)*
3. c *(35)*
4. f *(37)*
5. b *(37, 39)*

True/False

B.

1. True *(27)*
2. False. A correctly conducted inspection *is as much a public fire education program* as it is a code enforcement program. *(27)*
3. False. *Try not to be critical* of the occupant, building architect, contractor, or any one else associated with the property. *(28)*
4. False. The fire inspector should *not accept any favors* from the occupant, but rather, kindly refuse such offers. *(28, 29)*
5. False. *Do not assume that all occupants are ignorant of the codes.* Instead, fully explain the problems that have been noted and offer solutions for their correction. *(29)*
6. True *(30)*
7. False. The highest-ranking employee may choose to *delegate the responsibility* for hosting the fire inspector. *(32)*
8. True *(33)*
9. True *(33)*
10. False. It is *not* important for the inspector to draw a field sketch to scale. Dimensions can be written on the sketch so that a scale drawing may be developed later. *(34)*
11. True *(40)*
12. False. The methods for cataloging and storing inspection files *will vary from jurisdiction to jurisdiction.* *(42)*

13. False. *Few departments have the resources to develop their own inspection data management systems,* so most use a program that has been designed by an outside firm. *(42)*

Multiple Choice

C.

1. c *(27)*		13. b *(32)*
2. a *(27)*		14. d *(33)*
3. d *(27)*		15. d *(33, 34)*
4. b *(28)*		16. a *(34)*
5. a *(28)*		17. b *(34)*
6. c *(29)*		18. b *(36)*
7. d *(29)*		19. c *(37)*
8. d *(29)*		20. a *(37)*
9. a *(31)*		21. d *(39)*
10. a *(31)*		22. c *(40)*
11. b *(31)*		23. c *(42)*
12. c *(32)*		

Identify

D. *(36)*

1, 5, 6, 8

E. *(41)*

2, 3, 6

Chapter 3   Answers

Matching

A.

1. r *(47)*
2. l *(47)*
3. n *(47)*
4. j *(47)*
5. d *(48)*
6. b *(48)*
7. q *(49)*
8. k *(49)*
9. h *(50)*
10. a *(50)*
11. i *(51)*
12. g *(51)*
13. t *(52)*
14. m *(56)*
15. c *(60)*
16. f *(67)*
17. s *(69)*
18. o *(70)*
19. e *(71)*

True/False

B.

1. True *(47)*
2. True *(50)*
3. False. A positive heat balance occurs *when heat is fed back to the fuel.* If heat is dissipated faster than it is generated, a negative heat balance is created. *(51)*
4. True *(55)*
5. True *(56)*
6. True *(56)*
7. False. Fire resistance is a structural assembly's capacity *to maintain its load-bearing ability under fire conditions. (57)*
8. True *(57)*
9. False. Tests for *window assemblies involve both fire endurance and hose stream tests. (58)*
10. False. The *higher* the flame spread rating, the *more hazardous* the material. *(59)*
11. False. Fire load ratings *do not* accurately indicate fire severity with combustibles that have a high heat of combustion. The effects of the design and contents of a building on fire growth and spread may be more accurately determined by using computerized fire modeling programs. *(61, 62)*
12. False. Rain tests determine *if rain adversely affects a roof's fire-retardant abilities. (63)*
13. True *(64)*
14. True *(64)*
15. True *(65)*
16. False. *A fire door assembly includes* the door, the door frame, *the door closing and latching hardware, and other accessories. (65)*
17. False. Fire doors that serve as part of an exit *may be latched but not locked* against egress. Fire doors must be closed during a fire to provide an effective barrier. *(66)*
18. True *(67)*
19. False. Venting a building involved in a fire *improves visibility* for occupants and firefighters. *(67)*
20. True *(70)*
21. True *(71)*
22. False. Borax, boric-acid solutions, diammonium phosphate, and ammonium sulfate *are not truly effective fire retardants and should be avoided.* In reality, these substances deplete the moisture content in tree, and in the long run, increase the fire risk posed by the tree. *(72)*

Multiple Choice

C.

1.	c *(47)*	24.	c *(59)*
2.	d *(47)*	25.	d *(59)*
3.	b *(47)*	26.	b *(60)*
4.	b *(48)*	27.	a *(60)*
5.	a *(48)*	28.	c *(60)*
6.	c *(48)*	29.	d *(61)*
7.	c *(49)*	30.	a *(62)*
8.	d *(49)*	31.	d *(63)*
9.	b *(50)*	32.	b *(64)*
10.	d *(50)*	33.	b *(64)*
11.	a *(51)*	34.	d *(64, 65)*
12.	c *(51)*	35.	c *(66)*
13.	c *(51)*	36.	a *(66)*
14.	b *(52)*	37.	d *(66, 67)*
15.	a *(53)*	38.	c *(67, 68)*
16.	c *(54)*	39.	a *(68)*
17.	d *(54)*	40.	b *(69)*
18.	b *(54)*	41.	b *(70)*
19.	d *(55)*	42.	d *(71)*
20.	a *(56)*	43.	c *(71)*
21.	b *(56)*	44.	d *(71)*
22.	c *(58)*	45.	c *(72)*
23.	a *(59)*	46.	a *(72)*

Identify

D.

1. Underwriters Laboratories, Inc. *(58)*
2. National Institute for Standards and Technology *(58)*
3. Factory Mutual System *(58)*
4. American Society for Testing and Materials *(59)*
5. U.S. National Bureau of Standards *(60)*

Chapter 4 Answers

Matching

A.

1.	d *(75)*	6.	f *(75)*
2.	j *(75)*	7.	b *(77)*
3.	i *(75)*	8.	g *(77, 78)*
4.	a *(75)*	9.	c *(78)*
5.	h *(75)*		

B.

1.	k *(87)*	7.	h *(90)*
2.	d *(87)*	8.	a *(92)*
3.	j *(88)*	9.	f *(94)*
4.	l *(89)*	10.	i *(94)*
5.	e *(89)*	11.	b *(94)*
6.	c *(89)*		

C.

1.	e *(96)*	5.	g *(97)*
2.	h *(96, 97)*	6.	f *(97)*
3.	i *(97)*	7.	b *(98)*
4.	a *(97)*	8.	d *(99)*

D.

1. a *(103)*
2. d *(104)*
3. b *(103)*

True/False

E.

1. False. Most electrical fires are caused by *arcs or overheating. (75)*
2. False. Currents of 100 to 200 milliamperes or more *may be lethal. (76)*
3. True *(77)*
4. False. Driveways between stacks in an open-storage lumberyard should be a *minimum of 15 feet (4.5 m) wide. (79)*
5. True *(81)*
6. False. Attempts to keep incompatible products away from each other are often futile. *(82)*
7. True *(84)*
8. False. *No* combustible materials should be stored in cutting or welding areas. *(85)*
9. False. A hot work program uses a form that requires the person *issuing* the permit to ensure adherence to all fire prevention safeguards. *(86)*
10. True *(87)*
11. False. Class 2 filters, when clean, *will burn moderately* and emit *moderate* quantities of smoke. *(90)*
12. True *(91)*
13. True *(93)*
14. True *(94)*
15. False. Wet chemical systems are *not* recommended for electrical fires *because the spray may act as a conductor. (95)*
16. True *(96)*
17. True *(97)*
18. False. Dipping processes should *never* be located near an egress area. *(98)*
19. False. Any tank over *500 gallons (2 000 L)* must have a bottom drain that will open automatically or manually in the event of a fire. *(99)*
20. False. The first explosion usually introduces additional particles to the air, which results in *larger and stronger explosions. (101)*
21. True *(102)*

22. False. Fine metal dust has an *enormous* explosive potential. *(103)*
23. True *(106)*
24. False. Fires in approved burn containers should not be within *25 feet (7.5 m)* of a structure. *(107)*
25. True *(108)*
26. True *(110)*

Multiple Choice

F.

1.	d *(75)*	29.	a *(92)*
2.	b *(76)*	30.	c *(93)*
3.	c *(76)*	31.	b *(93)*
4.	d *(76)*	32.	b *(93)*
5.	c *(77)*	33.	a *(94)*
6.	a *(78)*	34.	d *(95)*
7.	c *(79)*	35.	c *(96)*
8.	a *(79)*	36.	d *(97)*
9.	d *(80)*	37.	b *(97)*
10.	b *(80)*	38.	a *(98, 99)*
11.	b *(80)*	39.	d *(99, 100)*
12.	c *(81)*	40.	b *(100)*
13.	c *(81)*	41.	d *(100)*
14.	d *(82)*	42.	d *(100)*
15.	c *(82)*	43.	d *(100)*
16.	a *(83)*	44.	c *(101)*
17.	d *(83)*	45.	a *(102)*
18.	a *(83)*	46.	a *(102)*
19.	c *(83, 84)*	47.	d *(102)*
20.	b *(84)*	48.	d *(103)*
21.	b *(85)*	49.	b *(103)*
22.	d *(85)*	50.	c *(104)*
23.	c *(85)*	51.	b *(106)*
24.	c *(86)*	52.	d *(107)*
25.	b *(86)*	53.	a *(107)*
26.	a *(87)*	54.	c *(108)*
27.	d *(88)*	55.	a *(108, 109)*
28.	c *(91)*	56.	d *(109, 110)*

Identify

G.

1. Oxy-fuel gas welding *(85)*
2. Heating, ventilating, and air conditioning *(86)*
3. Liquefied petroleum gas *(108)*
4. Compressed natural gas *(108)*
5. Hazardous production materials *(109, 110)*

H. *(100)*

2, 5, 6

I. *(104–106)*

2, 4, 5, 7, 9

J. *(110, 111)*

1, 5, 7

Chapter 5   Answers

Matching

A.

1. f *(115)*
2. d *(115)*
3. g *(116, 118)*
4. a *(116–118)*
5. h *(117–119)*
6. e *(117–119)*
7. i *(117–120)*
8. b *(118)*

True/False

B.

1. False. The only major difference between NFPA Type I and NFPA Type II construction ratings is that the degree of fire resistance is *lower in Type II.* *(116)*
2. True *(118)*
3. True *(119)*
4. False. UBC Type IV buildings *must have permanent* partitions. *(119)*
5. True *(120)*
6. False. The primary difference between SBC Type I/II and Type IV buildings is the *required degree of fire resistance for each of the structural elements. (120)*
7. True *(120, 122)*
8. False. Educational occupancies include all buildings used for educational purposes *up through the 12th grade by six or more persons, 4 hours per day or more than 12 hours per week. (123)*
9. False. If a building has two different uses in distinctly different areas, *each area should be required to meet the applicable requirements for that type of occupancy.* *(125)*
10. True *(125)*

Multiple Choice

C.

1. b (115)
2. a *(115)*
3. d *(116)*
4. c *(116)*
5. c *(117)*
6. a *(116)*
7. b *(117)*
8. d *(117)*
9. d *(118)*
10. c *(118)*
11. a *(118)*
12. a *(118)*
13. c *(120)*
14. b *(120)*
15. c *(120)*
16. d *(120)*
17. b *(122)*
18. a *(123)*
19. b *(123)*
20. a *(123)*
21. c *(124)*
22. d *(124)*
23. b *(124)*
24. a *(124)*

Identify

D.

1. National Fire Protection Association *(115)*
2. Building Officials & Code Administrators International, Inc. *(115)*
3. International Conference of Building Officials *(115)*
4. *Uniform Building Code™ (115)*
5. Southern Building Code Congress International, Inc. *(115)*
6. *Standard Building Code© (115)*
7. Board for Coordination of Model Codes *(125)*

Classify

E.

1. Group A-1 *(126)*
2. Group A-4 *(126)*
3. Group A-5 *(126)*
4. Group E *(123, 127)*

5. Group F-1 *(127)*
6. Group F-2 *(127)*
7. Group H-4 *(128)*
8. Group I-2 *(129)*
9. Group M *(124, 130)*
10. Group R-2 *(130)*
11. Group R-4 *(130)*
12. Group S-1 *(131)*
13. Group S-2 *(131)*
14. Group U *(132)*

F.

1. Group A-2*(126)*
2. Group E *(123, 127)*
3. Group F *(124, 127)*
4. Group H-1 *(128)*
5. Group H-3 *(128)*
6. Group I Restrained *(129)*
7. Group M *(124, 130)*
8. Group R-3*(130)*
9. Group S-1 *(131)*
10. Group S-2 *(131)*
11. No classification *(132)*

G.

1. Group A, Division 2.1 *(126)*
2. Group A, Division 4 *(127)*
3. Group E, Division 3 *(127)*
4. Group F, Division 1 *(127)*
5. Group F, Division 2 *(127)*
6. Group H, Division 2 *(128)*
7. Group H, Division 3 *(128)*
8. Group H, Division 6 *(128)*
9. Group I, Division 1.1 *(129, 130)*
10. Group R, Division 3 *(131)*
11. Group U, Division 1 *(132)*
12. Group S, Division 3 *(131)*
13. Group S, Division 2 *(131)*
14. Group U, Division 2 *(132)*

Chapter 6 Answers

Matching

A.

1. h *(138)*	7. i *(142)*
2. e *(138)*	8. m *(142)*
3 j *(138)*	9. k *(143)*
4. b *(138)*	10. a *(145)*
5. c *(139)*	11. l *(154)*
6. f *(141)*	12. g *(154)*

True/False

B.

1. True *(136)*
2. True *(137)*
3. False. All the codes specify that a public way should have a width and height of *at least 10 feet (3 m)*. *(138)*
4. False. In general, doors should open in the *direction of travel to the exit. (140)*
5. False. Panic hardware *is not required* for all exit doors. Panic hardware is required only under specific occupancy classifications. *(140)*
6. True *(142, 143)*
7. False. *Exit passageways* have *stricter construction protection requirements* than exit access corridors. *(143)*
8. True *(143)*
9. True *(144)*
10. True *(153)*
11. False. Inmates should be released by a reliable means, *whether it be electrical, mechanical, or by master keys. (155)*
12. True *(155)*

13. False. Exit drills at hotels and motels *must be held frequently to keep all personnel familiar with the evacuation plan and their assigned duties. (155)*

Multiple Choice

C.

1. a *(135)*	17. d *(144, 145)*
2. c *(136)*	18. b *(150)*
3. d *(136)*	19. a *(150)*
4. d *(136)*	20. d *(151)*
5. b *(137)*	21. c *(152)*
6. a *(137)*	22. b *(152)*
7. b *(138)*	23. a *(152)*
8. c *(138)*	24. b *(153)*
9. d *(140)*	25. d *(153)*
10. b *(141)*	26. c *(153)*
11. a *(141)*	27. d *(153)*
12. c *(141, 142)*	28. d *(154)*
13. c *(142)*	29. a *(154)*
14. b *(143)*	30. c *(154)*
15. b *(143)*	31. b *(155)*
16. a *(143, 144)*	

Identify

D.

1. Authority having jurisdiction *(137)*
2. Americans with Disabilities Act *(142)*

E.

1, 2, 4, 6, 7, 9 *(155)*

Chapter 7 Answers

Matching

A.

1. l *(160)*
2. d *(160)*
3. g *(161)*
4. i *(161)*
5. a *(162)*
6. e *(162)*
7. j *(163)*
8. h *(163)*
9. k *(169)*
10. f *(172)*
11. b *(178)*

B.

1. j *(192)*
2. c *(193)*
3. a *(193)*
4. f *(193)*
5. i *(194)*
6. k *(195)*
7. b *(196)*
8. e *(198)*
9. h *(199)*
10. d *(202)*

C.

1. l *(205)*
2. c *(206)*
3. f *(206)*
4. g *(206)*
5. b *(206, 207)*
6. h *(207)*
7. d *(207)*
8. i *(207)*
9. k *(207)*
10. a *(213)*
11. j *(218)*

True/False

D.

1. True *(161, 162)*
2. False. The lower the RTI value, the faster a sprinkler device operates. (162)
3. True *(164)*
4. False. Competent plant personnel or a contracted sprinkler company should perform routine system maintenance. *(164)*
5. True *(168)*
6. False. NFPA 13D requires a 13 gpm (49 L/min) water supply for each sprinkler if there are two or more. *(181)*
7. True *(182)*
8. True *(182, 183)*
9. False. Standpipe systems cannot take the place of automatic sprinklers. Automatic sprinklers are still the most effective method of fire protection in high-rise structures. *(186)*
10. True *(186)*
11. True *(186, 187)*
12. True *(188)*
13. False. Unlike pumps on fire apparatus, fixed fire pumps are not rated at a particular pressure. *(192)*
14. False. When an electric horizontal pump is tested, the measured voltage should never be more than 5 percent below or 10 percent above the rated voltage. *(203)*
15. True *(207)*
16. False. Sedimentations are found chiefly in the bottom of water mains; encrustations can form around the entire inside wall of the water main. *(210)*
17. True *(217)*
18. False. In general, when flow testing a single hydrant, the test hydrant should be between the flow hydrant and the water supply source. The flow hydrant should be downstream from the test hydrant. *(218)*

7

E.

1.	b *(159)*	22.	c *(187)*
2.	c *(159)*	23.	a *(188)*
3.	a *(160)*	24.	b *(189)*
4.	d *(162)*	25.	c *(190, 191)*
5.	c *(164)*	26.	c *(191)*
6.	a *(167)*	27.	d *(192)*
7.	c *(167)*	28.	a *(194)*
8.	d *(167)*	29.	d *(194)*
9.	b *(168)*	30.	c *(195)*
10.	d *(172)*	31.	c *(197)*
11.	a *(172)*	32.	b *(198)*
12.	c *(173)*	33.	b *(202)*
13.	b *(175)*	34.	a *(208)*
14.	d *(175)*	35.	c *(208)*
15.	b *(179)*	36.	c *(209)*
16.	d *(179)*	37.	d *(209)*
17.	b *(180)*	38.	b *(210)*
18.	c *(181)*	39.	a *(210)*
19.	a *(181)*	40.	c *(213)*
20.	d *(183)*	41.	c *(214)*
21.	d *(184)*	42.	b *(217)*

Identify

F.

1. Response time index *(162)*
2. Outside screw and yoke *(163)*
3. Post indicator valve *(163)*
4. Wall post indicator valve *(163)*
5. Post indicator valve assembly *(163)*

G. *(174, 175)*

2, 3, 4, 5

Chapter 8   Answers

Matching

A.

1. f *(234)*
2. h *(235)*
3. d *(236)*
4. a *(236)*
5. c *(237)*
6. e *(237, 238)*
7. i *(238)*
8. b *(238)*

B.

1. e *(249)*
2. c *(251)*
3. a *(252)*
4. f *(255)*
5. g *(258)*
6. h *(258)*
7. b *(258)*
8. i *(258)*

C.

1. c *(266))*
2. k *(266)*
3. h *(267)*
4. g *(270)*
5. b *(270)*
6. j *(271)*
7. i *(271)*
8. f *(271)*
9. a *(271)*
10. e *(272)*

D.

1. l *(273)*
2. e *(274)*
3. a *(274)*
4. g *(274)*
5. d *(275)*
6. j *(276)*
7. h *(276)*
8. b *(278)*
9. k *(279)*
10. i *(281)*
11. c *(281)*

True/False

E.

1. False. Class C extinguishers *have no numerical ratings. (232)*
2. False. *Limited production of new halon agents continues* because there are some exceptions to the ban. *(236)*
3. False. *None* of the commercially available agents are a *"pound-for-pound"* (0.5 kg) *replacement* for halon. *(236)*
4. True *(237)*
5. True *(237)*
6. False. Pump-operated extinguishers' *primary advantage is that they can be filled from any available water source in the course of extinguishing a fire. (239)*
7. True *(244)*
8. True *(244)*
9. True *(247)*
10. False. A specialized extinguishing system *must extinguish* the fire to be successful. *(249)*
11. True *(250, 251)*
12. False. There is *no standard container color* for halon storage tanks. *(252)*
13. True *(254)*
14. True *(257)*
15. True *(265)*
16. False. Discharging AFFF through standard sprinklers rather than foam sprinklers *tends to improve* the spray and penetration *because the AFFF has a greater velocity.* *(265)*
17. True *(268, 269)*
18. False. *Depending on local preferences,* the fire department may allow entities *other than itself* to monitor the remote station. *(271)*
19. True *(278)*

20. False. Occupants will have to test systems on their own most of the time and document the results. *At intervals specified by the fire department, fire department personnel will also witness the tests. (282)*

Multiple Choice

F.

1.	c *(230, 231)*	25.	c *(251)*
2.	d *(231)*	26.	d *(251)*
3.	b *(230)*	27.	d *(253)*
4.	b *(231, 232)*	28.	c *(253)*
5.	b *(234)*	29.	c *(256)*
6.	d *(234)*	30.	b *(257)*
7.	c *(234)*	31.	c *(259)*
8.	d *(234)*	32.	d *(260)*
9.	c *(235)*	33.	b *(260)*
10.	b *(235)*	34.	a *(261)*
11.	b *(237)*	35.	c *(261)*
12.	d *(237)*	36.	a *(263)*
13.	d *(237)*	37.	b *(264)*
14.	c *(239, 240)*	38.	c *(265)*
15.	a *(242)*	39.	d *(265)*
16.	c *(243)*	40.	c *(267)*
17.	b *(243)*	41.	b *(267)*
18.	d *(243)*	42.	b *(272, 273)*
19.	d *(245)*	43.	a *(280)*
20.	c *(246)*	44.	c *(281)*
21.	a *(247)*	45.	a *(281)*
22.	b *(248)*	46.	b *(282)*
23.	a *(248)*	47.	d *(283)*
24.	b *(251)*		

Identify

G.

1. Aqueous film forming foam *(235)*
2. Air-pressurized water *(238)*
3. Pull, aim, squeeze, sweep *(242)*
4. Film forming fluoroprotein foam *(259)*
5. Ultraviolet *(279)*
6. Infrared *(279)*

H. *(281–285)*

1, 2, 6, 7, 11, 12, 13

Chapter 9   Answers

Matching

A.

1. f *(290)*
2. i *(290)*
3. c *(292)*
4. b *(292)*
5. d *(292)*
6. h *(292)*
7. g *(294)*
8. j *(294)*
9. e *(295)*

True/False

B.

1. True *(290)*
2. False. A *site plan* may show slopes that might affect placement of fire department ground ladders. *(292)*
3. False. The left elevation is on the left side of the building *when viewed from the front. (294)*
4. False. Fire inspectors *should note* whether *intersections are looped and provide adequate turning radii* and whether *surrounding grades are too steep for ladder deployment. (295)*
5. True *(297)*
6. True *(298)*
7. False. *Codes often permit less stringent requirements* for interior finishes when automatic sprinklers are provided. *(299)*
8. False. Exit illumination must be continuous *during the time the building is occupied. (302)*
9. True *(304)*
10. False. Detailed information about fire alarm and communications is *not* usually provided on general construction plans. *Separate plans for these systems must be submitted and approved. (306)*
11. True *(306)*

Multiple Choice

C.

1. d *(289)*
2. b *(289, 290)*
3. a *(290)*
4. d *(290)*
5. c *(290)*
6. a *(290)*
7. b *(292)*
8. a *(292)*
9. d *(292)*
10. c *(292)*
11. b *(294)*
12. d *(295)*
13. d *(295)*
14. c *(297)*
15. a *(298)*
16. d *(298)*
17. b *(299)*
18. d *(301)*
19. b *(302, 304)*
20. d *(304)*
21. a *(306)*
22. c *(307)*
23. b *(307)*

Identify

D.

1. Computer-aided drawing *(290)*
2. Heating, ventilating, and air-conditioning *(299)*

137

Chapter 10 Answers

Matching

A. *(312)*

1.	e	5.	a
2.	h	6.	f
3.	c	7.	d
4.	g	8.	b

B.

1.	i *(315)*	6.	c *(314)*
2.	d *(314)*	7.	a *(312)*
3.	j *(315)*	8.	f *(314)*
4.	b *(312)*	9.	h *(315)*
5.	g *(314)*		

C.

1.	e *(318, 319)*	5.	f *(319)*
2.	b *(318)*	6.	g *(322)*
3.	h *(322)*	7.	c *(318)*
4.	d *(318)*		

D.

1.	h *(322)*	5.	f *(322, 323)*
2.	e *(322)*	6.	d *(323)*
3.	a *(322)*	7.	b *(323)*
4.	g *(322)*		

True/False

E.

1. True *(317)*
2. False. There is *no set format* for an MSDS, but each sheet has eight sections that contain specified information. *(318)*
3. False. An MSDS from the United States *is not* acceptable in Canada *because they are slightly different*. *(322)*
4. True *(322)*
5. True *(325)*
6. False. These distances are set forth by *local building codes, fire codes, planning codes, or the authority having jurisdiction*. *(327)*
7. False. Fusible vents on portable tanks must be designed to operate at a temperature *not exceeding 300°F (149°C)*. *(334)*
8. True *(338)*

Multiple Choice

F.

1.	d *(315)*	14.	d *(325)*
2.	a *(316)*	15.	c *(326)*
3.	d *(317)*	16.	a *(326)*
4.	b *(317)*	17.	d *(327)*
5.	a *(317)*	18.	d *(327)*
6.	c *(317)*	19.	c *(329)*
7.	c *(317)*	20.	d *(330)*
8.	c *(318, 319)*	21.	c *(330)*
9.	c *(319)*	22.	d *(331)*
10.	d *(318)*	23.	a *(331)*
11.	a *(322)*	24.	b *(332)*
12.	d *(322)*	25.	b *(334)*
13.	b *(325)*	26.	b *(339)*

Identify

G.

1. U.S. Department of Transportation *(311)*
2. Material Safety Data Sheet *(312)*
3. Other regulated materials *(312)*
4. *North American Emergency Response Guidebook (315)*
5. Chemical Abstract Service *(318)*
6. Lower flammable limit *(323)*
7. Upper flammable limit *(324)*
8. Boiling Liquid Expanding Vapor Explosion *(324)*
9. Vapor recovery system *(339)*
10. Environmental Protection Agency *(339)*

Chapter 11 Answers

Matching

A.

1. d *(347)*
2. c *(347)*
3. b *(347)*
4. a *(346)*

B.

1. f *(357)*
2. d *(357)*
3. g *(358)*
4. j *(358)*
5. a *(359)*
6. c *(361)*
7. h *(361)*
8. i *(361)*

True/False

C.

1. False. In their normal storage and transport states, explosives *are relatively stable and can be handled safely. (345)*
2. False. Gases with a vapor density greater than 1.0 *will fall* and can be expected to *lay near the ground or lowest point* of a room, structure, or area. *(349)*
3. True *(353)*
4. True *(355)*
5. False. *Containers may be filled with a liquid (oil or water) to prevent a reaction from occurring. (357)*
6. False. LD_{50} refers to the *ingested dose* of a substance that was lethal to 50 percent or more of the animals tested when they *swallowed or ate the substance. (359)*
7. False. Inspectors *should ask to see* verification that employees are adequately trained to work with toxic materials. *(359)*
8. True *(360)*

9. False. The alarm panel should be stored in an area of the facility where radioactive materials *are not stored or handled. (362)*

Multiple Choice

D.

1. c *(346)*
2. b *(346)*
3. b *(348)*
4. c *(348)*
5. d *(348)*
6. a *(349)*
7. d *(349)*
8. a *(350)*
9. b *(350)*
10. b *(351)*
11. d *(351)*
12. a *(352)*
13. c *(351, 352)*
14. d *(353)*
15. d *(353)*
16. b *(354)*
17. a *(354)*
18. a *(355)*
19. b *(359)*
20. c *(360)*
21. b *(361)*
22. a *(363)*
23. b *(363)*
24. d *(363)*
25. c *(363)*

Identify

E.

1. Bureau of Alcohol, Tobacco, and Firearms *(346)*
2. U.S. Department of Transportation *(348)*
3. Methyl-ethyl-propane *(349)*
4. Methyl-acetylene-propane *(349)*
5. Canadian Transport Commission *(350)*
6. Code of Federal Regulations *(355)*
7. Threshold limit value *(358)*
8. Radiation Safety Officer *(361)*
9. Nuclear Regulatory Commission *(362)*

F. *(355–357)*

2, 4, 5, 7, 8

Notes

Notes